Creo Parametric 4.0 Advanced Tutorial

Roger Toogood, Ph.D., P. Eng.

Publications

SDC Publications

P.O. Box 1334
Mission, KS 66222
913-262-2664
www.SDCpublications.com
Publisher: Stephen Schroff

ISBN-13: 978-1-63057-097-2
ISBN-10: 1-63057-097-4

Printed and bound in the United States of America.

Creo Parametric 4.0
Advanced Tutorial

Preface

The purpose of this tutorial is to introduce users to some of the more advanced features, commands, and functions in Creo Parametric. The presentation and material covered has been updated for Creo 4.0, although much of the material still applies to Creo 1.0/2.0/3.0. The style and approach of the previous **Advanced Tutorial** have been maintained. Each lesson concentrates on a few of the major topics and the text attempts to explain the "Why's" of the commands in addition to a concise step-by-step description of new command sequences. Familiarity with the basic operation of Creo is assumed, and material presented in the previous **Tutorial** is taken for granted. In a few places in this **Advanced Tutorial**, those commands are reviewed to put the discussion in context with the extensions and options presented here.

The material covered represents an overview of what are felt to be commonly used and important functions. These include customization of the working environment, advanced feature creation (sweeps, round sets, draft and tweaks, UDFs, patterns and family tables), layers, Pro/PROGRAM, and some advanced drawing and assembly functions. The **Adfvanced Tutorial** is not exhaustive, and there are many areas and options that could not be included due to space limitations. Nonetheless, it is hoped that the selection of the material presented here will satisfy a broad range of users wishing to expand their understanding of the program and learn additional features on their own.

The **Advanced Tutorial** consists of eight lessons. Each lesson should be studied using a live Creo session in order to fully experience and absorb the material. Most of the material in the lessons is reinforced by repetition, on the grounds that "practice makes perfect." Each lesson should take from 2 to 3 hours to complete (Lesson #8 slightly longer). Additional exploration of the program is strongly encouraged throughout.

This **Advanced Tutorial** has been written and tested using Creo Parametric 4.0 Build F000 running under Windows 7 Professional.

About the Project

A continuing theme through the lessons is the creation of parts for a medium-sized modeling project. Project parts are given at the end of each lesson that utilize functions presented in that lesson. Final assembly is performed in the last lesson. The project consists of a small three-wheeled utility cart. The entire cart project (part creation and final assembly) should take from 20 to 30 hours to complete (not including drawings). A **Quick Reference** chart for all the project parts is included immediately after the Table of Contents (page ix). Mark this page!

In today's litigious culture, it is unfortunately necessary to mention that the cart design is purely for the demonstration of modeling techniques in Creo. The cart design has been developed entirely by the author and any resemblance to an existing device is unintentional and coincidental. In many ways, the design presented here is incomplete. The cart and its components have not been subject to any engineering analysis, for example to determine its load carrying capacity or safety factors. Its suitability to any particular task or manufacturing process is therefore unwarranted and the author cannot take responsibility for the performance of an actual physical embodiment of the model.

Acknowledgments

As always, this Tutorial would not have been possible without the support and patience of my family who have tolerated my determination to complete the work to the exclusion of other important activities. So to Elaine, Kate, and Jenny: a great big THANKS!

Stephen Schroff at SDC has also been continually supportive and enthusiastic about this and other projects. A better publisher could not be imagined. He is aided by a very capable crew at SDC - thanks to all but especially Zach, Karla, and Megan.

As always, comments and suggestions are welcome and can be sent to <roger.toogood@ualberta.ca>.

I hope you enjoy the Tutorial.

RWT
Edmonton, Alberta
17 April 2017

TABLE OF CONTENTS

Lesson 1 : Customization Tools and Project Introduction

Lesson 2 : Helical Sweeps and Variable Section Sweeps

Lesson 3 : Advanced Rounds, Drafts and Tweaks

Lesson 4 : Patterns and Family Tables

Lesson 5 : User Defined Features (UDF's)

Lesson 6 : Pro/PROGRAM and Layers

Lesson 7 : Advanced Drawing Functions

Lesson 8 : Advanced Assembly

PROJECT PARTS QUICK REFERENCE

Part Name	Description	Page
arm_brack	bracket on horizontal side tube for suspension arm	1 - 21
arm_lower	lower wheel suspension arm	2 - 29
arm_upper	upper wheel suspension arm	2 - 28
arm_vbrack	bracket on vertical side tube for suspension arm	1 - 21
cargo	cargo bin	3 - 29
fram_low_rgt	lower frame tube member on right side	2 - 31
fram_upp_rgt	upper frame tube member on right side	2 - 31
frame_front	front frame member	7 - 40
front_axle	front wheel axle	6 - 23
front_pillar	front wheel assembly pillar	7 - 39
front_spr_plate	circular plate to support front spring	1 - 20
front_spring	front wheel spring	2 - 28
front_wheel	front wheel	6 - 23
front_wheel_brack	front wheel mounting bracket	6 - 24
handle	front handle	5 - 24
handle_pin	axle pin for front handle	1 - 20
hex_bolt	generic part for hexagonal shoulder bolt	4 - 34
hubcap	wheel hubcap	3 - 28
lugnut	nut to attach wheel to axle	3 - 28
mount	side wheel axle mounting plate	5 - 24
pillar_cap	cap on top of front wheel assembly pillar	7 - 39
spring	main side wheel suspension spring	3 - 31
stud	original part to define stud UDF	5 - 26
tubing	generic part for square tubing	4 - 33
wheel	main side wheel	4 - 33
wheel_axle	side wheel axle	5 - 26

This page left blank.

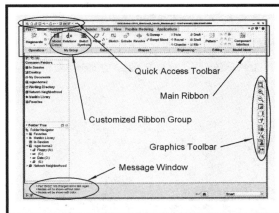

Lesson 1

*Customization Tools
and
Project Introduction*

Synopsis:

Configuration settings; customizing the screen toolbars and menus; mapkeys; part templates; introduction to the project

Overview

This lesson will introduce tools for customizing your configuration and working environment and show you how to create some useful shortcuts for accessing Creo commands. The intent of these tools is to let you set up the interface to suit your own preferences so that you can work most comfortably and efficiently. The major customization tool is the use of one or more configuration files (default file *config.pro*). The lesson also includes managing and creating your own custom toolbars and mapkeys. We'll also see how you can create your own part templates. The major project used in this tutorial is introduced and the first four parts are presented.

Configuration Settings

Launch Creo, or if it is already up erase everything currently in session and set your working directory to your normal start-up directory. You should now be looking at the **Home** ribbon, with nothing loaded.

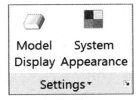

Figure 1 The **Settings** group in the **Home** ribbon

The **Settings** group in the **Home** ribbon is shown in Figure 1. This group is available only in the **Home** ribbon. Select either of these commands (*Model Display*, *System Appearance*), or the arrow in the lower right corner, and you will see that each is going directly to one of the option areas in the same **Creo Parametric Options** window. At all other times (i.e. when you have a part or assembly loaded), this window is available by selecting *File ➤ Options*.

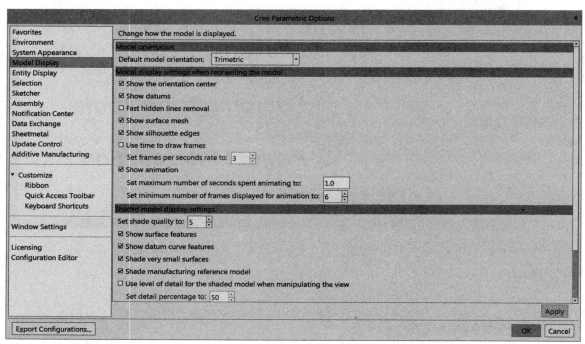

Figure 2 The **Model Display** panel in the **Creo Parametric Options** window

The **Options** window is the main location for most of the customization tools. Select the **Model Display** category as in Figure 2. The panel on the right shows all the settings related to the chosen category. Have a look at some of the other categories and options available.

Any setting changes made will be valid for the current session unless you explicitly save them. That is done using the **Export Configurations** button at the bottom of the window. In addition, at the bottom of some of the panels is an **Import/Export** button (that also allows you to read in a file of previously created settings).

If you have made settings changes but have not explicitly saved them, when you select **OK** in the **Options** window, you will be asked if you want to save the settings. A "yes" will open the dialog window shown in Figure 3. The default name and location for the configuration settings file is *config.pro* in the current working directory. If you want the settings to "stick" (be active the next time you launch Creo), this is exactly where you want them saved. If you don't want the settings read in at start-up, but do want them available later for a specific project, you can save the configuration file anywhere. It can then be loaded at any time using the **Import** function.

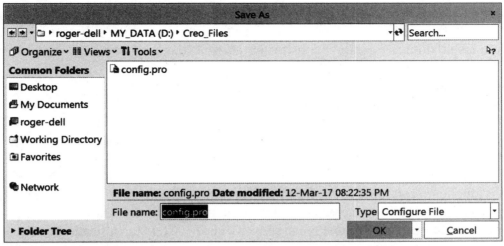

Figure 3 Saving the configuration settings

Now let's have a closer look at the contents of the *config.pro* file itself.

Configuration Files (*config.pro*)

As mentioned above, the most important configuration setting file is a special file called *config.pro* that is automatically read from the startup directory when you launch a new session. You can also read in (and/or change) additional configuration settings at any time during a session. For example, you may want to have one group of settings for one project you are working on, and another group for a different project that you switch to during a single session. In this tutorial, we will deal only with the use of the single configuration file, *config.pro*, loaded at start-up.

Several copies of *config.pro* might exist on your system, and they are read in the following order when Creo is launched:

♦ *config.sup* - this is a protected system file which is read by all networked users but is not usually available for modification by users. Your system administrator has control of this file.
♦ Creo loadpoint/text directory - the *config.pro* file stored here is read by all users and would usually contain common settings determined by the system administrator such as search paths, formats, libraries, and so on. This file cannot normally be altered by individual users.
♦ user home directory - unique for each user (Unix).
♦ startup directory - the working directory when Creo starts up[1]. This is highlighted in the Navigator on start-up.

---

[1] In Windows, right click on the Creo icon on the desktop (if it exists), select **Properties ➤ ShortCut** and examine the **Start In** text entry field.

Settings made in the first copy (*config.sup*) are locked and cannot be overridden by users. This is handy for making configuration settings to be applied universally across all users at a Creo installation site (search paths for libraries or drawing templates, for instance).

During the installation procedure for the software, a copy of a stripped-down *config.pro* file was created in the **../text** directory of the software loadpoint. The contents of this file depend on which system of units was chosen (English or metric) during software installation. If metric was chosen, a number of options are set to reflect ISO standards. If English was chosen, these options are set to ASME and ANSI standards.

An individual user can modify entries in the last two copies of *config.pro* to suit their own requirements. If the same entry appears more than once, the last entry encountered in the start-up sequence is the one the system will use. After start-up, additional configuration settings can be read in at any time. These might be used to create a configuration unique to a special project, or perhaps a special type of modeling (sheetmetal, for example). Be aware that when a new configuration file is read in, some options may not take effect until Creo is restarted. This is discussed more a bit later.

The Configuration File Editor

You can access your current configuration file using

File ➤ Options ➤ Configuration Editor

This brings up the **Options** window with the existing *config.pro* contents shown at the right. See Figure 4. The ***Show*** field indicates the source of the currently listed options.

Click the ***Show*** pull-down at the right, and select ***All Options***. A complete list of all the Creo configuration options will appear. The first column shows its name, and the second column shows its current value. A value with an asterisk indicates a default value. The third column indicates its status (a solid green circle means the option has been read from the *config* file, rather than the system default), followed by a one-line description in the fourth column. Note that you can resize the column widths by dragging on the vertical column separator bars at the top of the display area.

Browse down through the list. There are a lot of options here (about 1000!). Note that the options are arranged alphabetically. This is because of the setting in the **Sort** pull-down menu in the top-left corner. Change this to ***By Category***. This rearranges the list of options to group them by function. For example, check out the settings available in the **Environment** and **Sketcher** groups. The list of options is a bit overwhelming. Fortunately, there are a couple of tools to help you find the setting you're looking for. Let's see how they work.

Open the **Show** list again and select **Current Session** and select ***Sort(Alphabetical)***. There will likely be some options listed, if not from a *config.pro* in your startup directory, then from the software loadpoint. You can see these by selecting from the options in the **Show** list. In that list, selecting ***Only Changed*** makes it easier to see new settings.

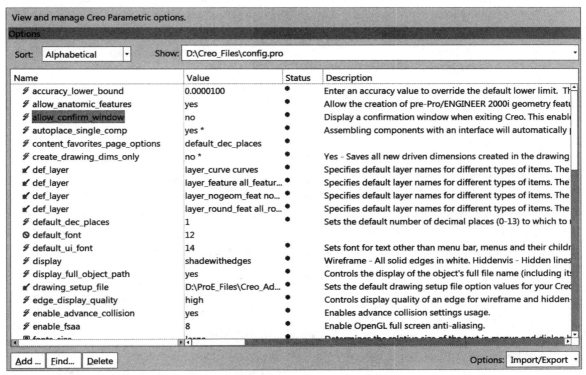

Figure 4 The *config.pro* file layout showing current settings

Adding Settings to *config.pro*

Let's create a couple of useful settings.
At the bottom of the **Options** window
are two buttons: *Add* and *Find*. If you
know the name of the option, you can
just select *Add* and type it in the top box
shown in Figure 5.

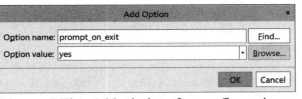

Figure 5 The *Add* window for config options

For new users, a useful setting is the following. In the name box enter the option name
prompt_on_exit. As you type this in, notice that Creo anticipates the rest of the text
based on the letters you have typed in. After typing enough characters (up to the "x" in
"exit" - it takes Creo that long to find out you don't want ***prompt_on_erase***), the rest of
the desired option will appear. In the pull-down list beside **Value**, select *Yes* (or just type
"y"). Note that the option name is not case sensitive and the default value is indicated by
an asterisk in the pull-down list. Now select the *OK* button. The entry now appears in
the data area. A bright green star in the **Status** column indicates that the option has been
defined but has not yet taken effect.

Now enter a display option. The default part display mode in the graphics window is
Shaded. Some people prefer to work in shaded mode with part edges shown - let's make
it the default on start-up. Once again, we will enter the configuration option name using
Add and pick the value from a drop-down list. The option name and value we want are

| display | shadewithedges |

Now select **OK** as before. Add the following option to control how datum planes are displayed during dynamic spinning with the mouse:

| **spin_with_part_entities** | yes |

Another common setting is the location of the Creo trail file. As you recall, the trail file contains a record of every command and mouse click during a Creo session. The default location for this is the start-up directory. Theoretically, trail files can be used to recover from disastrous crashes, but this is a tricky operation. Most people just delete them. It is handy, therefore, to collect trail files in a single directory, where they can be easily removed later. There is an option for setting the location of this directory. Suppose we don't know the configuration option's specific name. Here is where a search function will come in handy.

At the bottom of the **Options** window, click the **Find** button (you may have noted that this is also available in the **Add** window, Figure 5). This brings up the **Find Option** window (Figure 6).

Type in the keyword **trail** and leave the default **Look in(ALL_CATEGORY)** (note the check box option to search the descriptions for the keyword; we don't need that in this case) then select **Find Now**.

Several possibilities come up. The option we want is listed as **trail_dir** - scroll the description to the right to confirm this. Select this option and then pick the **Browse** button at the bottom to identify a suitable location on your system for the

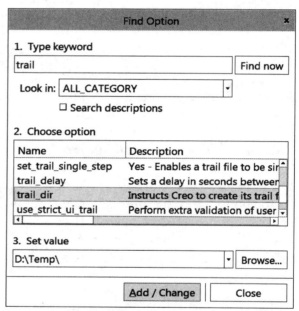

Figure 6 Using **Find** to locate an option

value, perhaps something like *c:\temp*. Then select **Add/Change**. The new entry appears in the **Options** window. In the **Find Option** window, select **Close**.

For some options, the value is numeric (e.g. setting a default tolerance, number of digits, or the color of entities on the screen). In these cases, you can enter the relevant number (or numbers separated by either spaces or commas).

For practice, enter the following options, either directly or using **Find**. (HINT: use the RMB pop-up menu anywhere in the option table and select **Add**.) The order that the configuration options are declared does not matter. Feel free to add new settings to your file (for search paths, libraries, default editors, default decimal places, import/export settings, and so on).

allow_anatomic features	yes
default_dec_places	1
display_full_object_path	yes
sketcher_starts_in_2D	yes
sketcher_lock_modified_dims	yes

In the Show pull-down list, select *Current Session*. Notice the icons in the first column beside the option names. These mean the following:

     ⚡ (lightning) - option takes effect immediately
     ✦ (wand) - option will take effect for the next object created
     ▣ (screen) - option will take effect the next time Creo is started
     ⊘ (illegal) - option is illegal or not recognized

If you are using a *config* file from a previous version of Creo you may see the red illegal symbol, which means that the option is no longer used.

Try to add an illegal option name. For example, in a previous release there was an option **sketcher_readme_alert**. Try to add that and set it to **Yes**. When you try to set a value for this, it will not be accepted (the *OK* button stays gray). Creo only recognizes valid option names! Thus, if you mistype or enter an invalid name, this is indicated by not being able to enter a value for it.

Saving Your *config.pro* Settings

The green stars in the **Status** column indicate the new settings have not taken effect yet. To store the settings we have just created, select either the *Export Configurations* button, or open the *Import/Export* menu of the **Options** window. In the Save As window that opens, type in the desired name for the file (or accept the default) - in this case *config.pro* and select *OK*.

Loading a Configuration File

You may have noticed (in the Import/Export menu), the Import Configuration command. Select that now and pick the desired file and then *Open*. These settings will be read in and added to the current ones but not activated immediately (note the green star). That will happen when you close the window with *OK*.

Deleting Configuration Options

Highlight one of the options (maybe an illegal one) and use the RMB pop-up to select *Delete*. Save the modified file with *Export*, then use *OK* to close the window.

Checking Your Configuration Options

Because some settings (mainly dealing with windows properties like fonts and colors) will not activate until Creo is restarted, many users will exit after making changes to their *config.pro* file and then restart, just to make sure the settings are doing what they are supposed to. Do that now. This is not quite so critical since the **Options** window shows you with the lightning/wand/screen icons whether an option is active. However, be aware of where Creo will look for the *config.pro* file on start-up, as discussed above. If you have saved *config.pro* in another directory than the one you normally start in, then move it before startup. On the other hand, if you have settings that you only want active when you are in a certain directory, keep a copy of *config.pro* there and load it once Creo has started and you have changed to the desired directory. To keep things simple, and until you have plenty of experience with changing the configuration settings, it is usually better to have only one copy of *config.pro* in your startup directory.

Note that it is probably easier to make some changes to the environment for a single session using *File* ➤ *Options*. Also, as is often the case when learning to use new computer tools, don't try anything too adventurous with *config.pro* in the middle of a critical part or assembly creation session - you never know when an unanticipated effect might clobber your work!

Customizing the Toolbars

For the following, you will have to load one of the parts you have previously made. Do that now.

Click the right mouse button on the Graphics toolbar. This brings up a menu like the one shown in Figure 7. This allows you to select which commands appear on the toolbar, where the toolbar is located, the size of the toolbar icons, or reset to the default toolbar.

The option *Propagate Customization* will keep the same toolbar across all modes of Creo (into Simulate, for example).

Now use the RMB pop-up on the Quick Access toolbar at the top. A pop-up will allow you to specify the location (relative to the ribbon) or remove it entirely, or modify the commands. If you select

Customize Quick Access Toolbar

you will be taken to the regular Creo Parametric Options window. Here you may select from any command in the left pane and *Add* it to the toolbar list on the right. You can change its position relative to the other commands using the up/down arrows at the right side of the window. Notice that there is a button that will reset to all default values. Also, be aware that some commands will

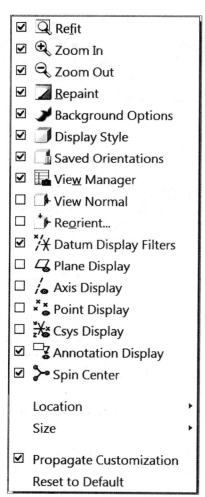

Figure 7 Customizing the **Graphics** toolbar

not be available unless you are in relevant modes (for example, the Model Colors command is not available unless a model is loaded). Close this window.

Customizing Ribbon Tabs and Groups

With the mouse on the ribbon, use the RMB and select *Customize the Ribbon*. This opens the usual **Options** window arranged as shown in Figure 8. The Creo commands are listed in the left pane, and the ribbon structure is shown in the right pane. Each of the tab titles and groups can be expanded to show the commands it contains.

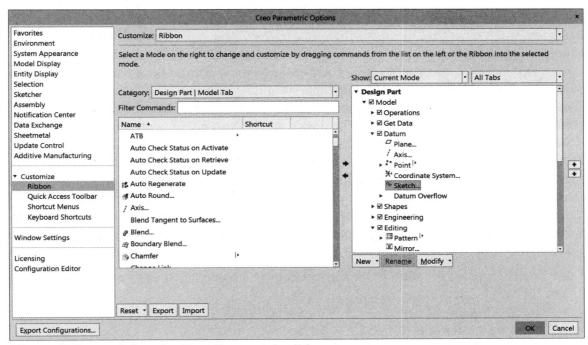

Figure 8 Customizing the Creo Parametric ribbons

We can do a lot to modify the contents and appearance of the ribbons. First, each tab and group has a checkbox to specify whether it should be displayed. Within each group, we can modify the appearance of the command icons. For example, select the *Mirror* command in the *Editing* group of the *Model* tab. Now you can select the *Modify* command at the bottom to change the size and appearance of the command. It is also possible to do this a bit more easily by moving the **Option** window out of the way and just selecting the button in the ribbon and using the RMB pop-up menu. Depending on how much space there is on your screen, you may find that the appearance of other commands in other groups may change (mainly the command label).

Select the **Model** tab, then using the RMB menu select *Add New Group*. We can specify a new name with *Rename*. Now with the new group selected, we can move commands from the left pane into the new group. This allows you to collect commands that you use often that are located on different ribbons. Create the group shown in Figure 9. You can remove commands, groups, and tabs using the RMB pop-up menu and selecting *Remove*.

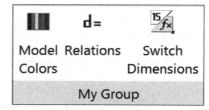

Figure 9 A customized ribbon group

As usual, to make these changes permanent, you must *Export* them.

Helpful Hint

It is tempting, if you are blessed with a lot of screen space, to over-populate the toolbars and ribbons by trying to arrange every commonly used command on the screen at once. Before you do that, you should work with Creo for a while. You will find that Creo will generally bring up the appropriate toolbars for your current program status automatically. For example, if you are in Sketcher, the Sketcher short-cut buttons will appear. Thus, adding these buttons permanently to any toolbar is unnecessary and the buttons will be grayed out when you are not in Sketcher anyway - you are introducing screen clutter with no benefit. Furthermore, many commands are readily available in the right-mouse pop-up menus. Since these are context sensitive, you will only have to choose from commands that are useful at that moment.

There is lots more information in the on-line help. A good place to start is to open the following path in the Help Center:

> **Fundamentals**
> > **Fundamentals**
> > > **Configuring Fundamentals**

You can also do a search in the Help pages for *config.pro*.

Keyboard Shortcuts - Hotkeys

The table at the right shows most of the default hotkeys. These are shortcuts to commonly used commands. As indicated, these are launched by holding down the **Ctrl** key while the keyboard letter is pressed. Most of these are mnemonic (indicated by the **_highlighted_** letter).

Key	Command
Ctrl - A	**A**ctivate window
Ctrl - C	**C**opy feature
Ctrl - D	**D**efault view orientation
Ctrl - G	re**G**enerate model
Ctrl - N	create **N**ew model
Ctrl - O	**O**pen existing model
Ctrl - P	**P**rint
Ctrl - R	**R**epaint graphics window
Ctrl - S	**S**ave file
Ctrl - V	paste feature
Ctrl - Y	redo
Ctrl - Z	undo

Keyboard Shortcuts - Mapkeys

A mapkey is a short sequence of keyboard key strokes or a function key that will launch one or a series of Creo commands. Since many simple commands are launched using toolbar buttons, mapkeys are typically used to start extended command sequences. Mapkeys are very similar in usage to macros that can be defined in other software packages. Mapkey definitions are contained/included in your *config.pro* file, so they are loaded at start-up. Mapkeys are meant to be used for extended command sequences that you use frequently.

The mapkey can be as long as you want; most users restrict mapkeys to only 2, or sometimes 3, lower case characters. This gives several hundred possible mapkey sequences - more than you can probably remember effectively. Creo constantly monitors the keyboard for input and will immediately execute a defined command sequence when its mapkey is detected. Single character mapkeys should be avoided due to the way that Creo processes keyboard input. If you have two mapkeys "v" and "vd", for example, the second mapkey would never execute since Creo will trap and execute the first one as soon as the "v" is pressed (and in any case, Creo has a lot of one-character mapkeys already, like "D" to create a datum plane. Look for these single letter commands in the button pop-ups that show when you hover the mouse over a command in the ribbon).

Ideally, you would like to have mapkeys that are very easy to remember, like "vd" (view default), or "rg" (regenerate). Furthermore, mapkeys are usually launched with your opposite hand from the mouse. Because it is common to only use two characters, it will take some planning to decide how you want to set up your definitions to use only a couple of easy-to-remember key strokes! The mapkey should be mnemonic but can't collide with other definitions. You don't want to have to remember that "qy" means "repaint the screen."

A practical limit on usable mapkeys is perhaps in the range of 10 to 20, although some "power users" can use many more. In previous versions of the software, some commands used to be several menus deep into the program, requiring numerous mouse clicks, so mapkeys made life a lot easier. However, with the development of the Creo interface, the need for dozens of mapkeys is diminishing - you can probably get by with just a few (or even none!).

Listing Current Mapkeys

To see a list of your current mapkeys (some may be defined in your *config.pro* file) select

> ***File > Options***
> ***Environment***
> ***Mapkeys Settings***

This dialog window (Figure 10) allows you to define and record, modify, delete, run, and save mapkeys. Note that each mapkey has a short **Name** and **Description**. The **Name** will be used on any short-cut button (described below), and the **Description** will appear in the message area in the main graphics window. Mapkeys that start with a "$" are function keys.

Note that mapkeys created using a previous release of Creo may differ in command syntax and it is likely that some mapkey definitions from previous releases will not function properly. However, mapkeys are easy enough to record.

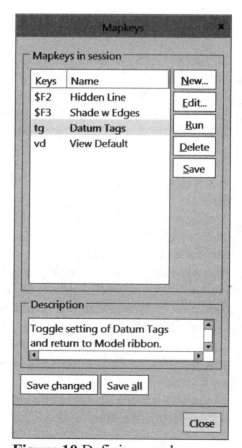

Figure 10 Defining mapkeys

In the following, it is assumed that you have no mapkeys defined as yet. If any of these tutorial mapkeys collide with existing mapkeys shown in the mapkeys list (Figure 10) or the built-in keyboard shortcuts, you can modify the keyboard sequence (for example, use "dv" instead of "vd") for the new mapkey.

Creating Mapkeys

New mapkeys are created as follows. We will create a very simple mapkey sequence "vd" that will reorient the view to the default orientation and, from wherever you are in the ribbons, return you to the **Model** ribbon. To set this up, you will have to bring in one of your previously created parts. Do that now. We will not be modifying the part.

Select the *New* button in the **Mapkeys** menu (Figure 10). The **Record Mapkey** dialog box shown in Figure 11 will open. Enter the data shown in the figure: key sequence, name, description. Now we record the command sequence while we execute the desired command sequence:

> *Record*

Click on the **View** ribbon tab and select *Standard Orientation*. Now click on the **Model** ribbon tab. Finally, in the **Record Mapkey** dialog window select

> *Stop > OK*

Every command and keystroke between *Record* and *Stop* is added to the mapkey. It's that easy! Spin the model with the middle mouse button and select any other ribbon. In the mapkeys window, highlight the new mapkey "vd" and select the *Run* button. You should end up in default view orientation back in the **Model** ribbon. It's a good idea to check your mapkey definitions now when it is easy to modify them.

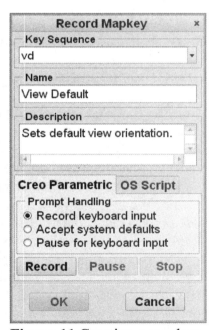

Figure 11 Creating a mapkey

As mentioned above, mapkey definitions are saved in a configuration file (as in *config.pro*). New mapkey definitions are appended to the end of the file. If you redefine a mapkey (or use a duplicate keystroke sequence), the definition closest to the bottom of the *config* file is the one that will be used. When saving a mapkey you can choose either *config.pro* or *current_session.pro*. There are three ways to save the mapkeys using the buttons in the **Mapkeys** window:

> *Save* - saves only the highlighted mapkey
> *Save Changed* - saves any mapkeys changed this session
> *Save All* - saves all mapkeys defined for the current session

Remember that if you save the mapkey in the *current_session.pro* or elsewhere, it will not be loaded automatically the next time you start Creo. To do that, you must explicitly save the mapkey definitions into the *config.pro* file. Select one of the three options and save our "vd" mapkey.

Also, be aware that if you save all mapkeys defined in the session, they are appended to the end of the *config.pro* file. If you do this excessively, the file can become quite large (and slow to load). You should occasionally edit the file with a text editor to remove the early duplicates.

Close the **Mapkeys** window. Minimize Creo and open *config.pro* using your system text editor. Scroll down to the bottom of the file to see the new lines that describe the mapkey - it will extend over several lines. It is possible to move the mapkey definitions elsewhere in the file, but for each definition these lines should never be separated since they are a continuation of the same sequence. It is possible, but probably not advisable, to try to edit the mapkey definitions manually - leave that to the power users! Exit your text editor and restore the Creo window.

Some final points about mapkeys: it is possible to set up the mapkey so that execution will pause to allow user input during the command sequence, either by picking on the screen or through the keyboard. Mapkeys can also call other mapkeys. You might like to experiment with these ideas on your own. The possibilities for customization are almost limitless! As mentioned earlier, with the "flattening" of the user interface in Creo, the need for mapkeys is greatly diminished. They are primarily used as shortcuts for frequently used sequences of multiple commands.

Creating a Customized Part Template

Most part files that you create contain many common elements such as datums, defined views, coordinate systems, parameters, and so on. Creating these from scratch for every new part that you start is tedious and inefficient. Prior to Release 2000i^2 a very handy model creation tool used the notion of a "start part" which contained these common elements. Users would then create a mapkey that would bring the part into session and then rename it. This made the creation of new parts very quick and efficient, with the added bonus that standard part setups could be employed.

This "start part" functionality has been built into the program using part templates. Several part templates are included with a standard Creo installation for solid and sheet metal parts in different systems of units. You may have a reason at some point to create your own template, which we will do here. We'll also set up the system so that your template becomes the default when you create a new part. Then you can immediately get on with the job of creating features. We will create the custom part template using one of the existing templates as a basis. This demonstrates that any part file can be used as a template, even ones containing existing solid features.

Select *File > New* (or Ctrl-N). Make sure the **Part** and **Solid** radio buttons are selected. Deselect the **Use Default Template** box, and enter a name *mytemplate*. Select *OK* and in the next window, select the **mmns_part_solid** template. Notice that it has two parameters (MODELED_BY and DESCRIPTION). Fill in the MODELED_BY parameter with your name or initials. Select *OK*.

Now we will customize the existing part. You may find that you don't often use the default part coordinate system, so delete that. If you do a lot of axisymmetric parts, you might like a default axis. Create one now by selecting (with Ctrl) the RIGHT and FRONT datums, then the *Axis* command. The part should look like Figure 12.

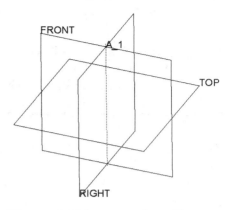

Figure 12 New template file *mytemplate.prt* default datums

Let's create a parameter for the part material. Go to the **Model Intent** group pulldown and select *Parameters*. The two existing parameters are listed. Click the add button ("+"), and create a new parameter **MATERIAL** and set the type to *String*.

We are finished with creating the part, so save it in your default working directory with the name **mytemplate.prt**. If you have write access, copy the part file to the Creo installation TEMPLATES directory, something like

<Creo Loadpoint>\common files\templates

This is the default directory where Creo will look for part templates. If you do not have write access to this directory, leave the part file in your working directory. You can rename the file to remove the version number if you want, so that it appears as **mytemplate.prt** rather than **mytemplate.prt.1**.

Setting the Default Part Template

We can tell Creo to use our new template as the default by setting an option in *config.pro*. Select

File > Options > Configuration Editor

and *Add* (or edit) the option **template_solidpart**. Set the value for the option by browsing to the template directory (or use the current working directory, wherever you have saved the template file) and selecting the part file **mytemplate.prt** we created above. Alternatively, you can specify the location in the system templates directory using (type this exactly as shown)

template_solidpart $PRO_DIRECTORY\templates\mytemplate.prt

When you exit the options window, the new setting should be stored in your default configuration file (*config.pro*).

Using the New Part Template

Close the current part window and select *Erase_Not_Displayed*. Select

File > New

Make sure the **Use Default Template** box is now checked, enter a name (like *test*), and select *OK*. A copy of your custom template (check for the axis and the material parameter) is now brought into session and given the name you specified.

Adding Mapkeys to Ribbons

Any Creo command or mapkey can be added to any of the existing ribbons and toolbars. Let's add a mapkey to the group **My Group** that we created in the **Model** ribbon earlier in this lesson. Select

File > Options > Customize Ribbon

In the dropdown **Category** list, select **Mapkeys**. Pick the *View Default* mapkey that we made earlier, and *Add* that to **My Group** in the **Model** ribbon group list at the right. Below the list select *Modify* to edit or change color of the button image, or choose from among several hundred existing icons to replace the happy face with something easier to remember. See Figure 13.

Now, of course, you will only see the **My Group** buttons when the **Model** ribbon is open. To make the mapkey available at any time, it needs to go in the Quick Access toolbar at the top. Select

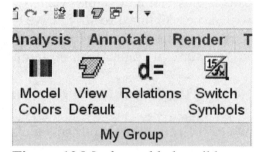

Figure 13 Mapkey added to ribbon group and Quick Access toolbar

File > Options
Quick Access Toolbar

Once again select **Mapkeys** in the pulldown list, then pick the **View Default** mapkey and move it to the list at the right with *Add*. You can position the icon on the toolbar using the up and down arrows at the far right. Close the **Options** window. Your settings will be saved for the next time you launch Creo.

Introduction to the Project

The assembly project to be completed in this tutorial involves the modeling and assembly of the three-wheeled utility cart shown in Figure 14. The cart contains 26 or so parts, many of which are repeated in the assembly. The total assembly has about 75 parts (mostly bolts!). We will use the techniques introduced in the lessons to model various parts of the cart as exercises at the end of each lesson. We will average about 4 parts per lesson, so you should get lots of practice! In the final lesson, we will assemble the cart using a number of advanced functions for dealing with assemblies. Try not to "jump the gun" on this assembly task, since the functions to be covered in the last lesson can really speed up your job of putting the cart together.

For your modeling exercises, the parts shown at the end of each lesson[2] will illustrate the critical dimensions. A figure will also be provided to show where the parts fit into the overall assembly. **Only the critical dimensions are shown on each part - you can use your judgement and creativity to determine the remaining dimensions.** In this regard, take note of the following:

♦ ALL UNITS ARE IN MILLIMETERS! We set up the default part template with this setting.
♦ Dimensions are usually multiples of 5mm. For instance, all the plate material and the wall of the cargo box are 5mm thick. The tubing is 25mm square.
♦ All holes and cylinders, unless otherwise dimensioned, are $\varphi 10$. This applies to bolt holes, pins, rods, and so on.
♦ All holes, unless otherwise dimensioned, are coaxial with cylindrical surfaces or located on symmetry planes.
♦ For some of the trickier parts, in addition to the figures showing the dimensions, there will be some discussion and hints to help you get going.

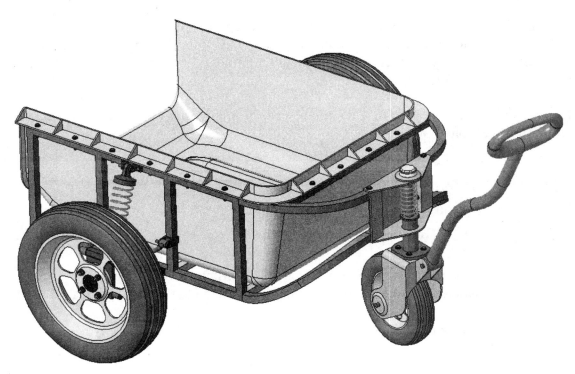

Figure 14 The assembly project - a three-wheeled utility cart

---

[2] A complete list of project parts is included in the preface (right after the Table of Contents) to this book.

When we get to the final assembly in Lesson 8, remember that it is an easy matter to modify dimensions of the various parts so that the assembly fits together. Don't be too concerned when you are modeling the parts if you have to guess at one or two dimensions. These can be modified later if the need arises.

When you are creating the parts, try to be aware of the design intent for the part and how it might eventually be placed in the assembly[3]. For example, if the part has one or more planes of symmetry, it is common practice to use the default datum planes for these. In the assembly, the *Align* constraint using these datum planes is an easy way to position the part (usually with another symmetric part).

Although a suggested part name is given, feel free to make up your own part names (although this might cause confusion in Lesson 8!). Remember that Creo is fussy about files that get renamed in isolation, or moved to another directory. If a part has been used in an assembly (or sub-assembly) or drawing, make sure the assembly or drawing is in session if you rename or move the part so that the related files can also be updated.

Summary

This lesson should have given you enough ideas and ammunition to allow you to customize the interface so that it will be most efficient for the type of work that you do. There are a surprising number of users who are unaware of the many options available in *config.pro*. Check them out!

In the next lesson we will look at functions directly involved in model creation. These are for the creation of sweeps.

Questions for Review

1. What is the name of the file containing your configuration settings?
2. How can you change the appearance of the command buttons on ribbons? What are the options?
3. When, and from where, are your configuration settings loaded? Why is there more than one location?
4. What happens if your configuration file contains multiple entries for the same option, each with different values?
5. How can you find out where your start-up directory is?

---

[3] You might like to look ahead to the last lesson to see what assembly constraints are used for each part.

6.   How can you create/edit/delete configuration settings?
7.   When do configuration settings become active?
8.   Is it possible to have more than one customized screen layout?
9.   How do you move the Graphics Toolbar in the graphics window?
10.   How do you add/delete buttons on the toolbars?
11.   How do you create your own ribbon group? What are the restrictions on commands that can be placed there?
12.   How do you add a button to the Quick Access toolbar?
13.   What is a mapkey?
14.   Why do you usually want to keep mapkey names short?
15.   How do you attach a mapkey to a keyboard function key?
16.   How do you create a new mapkey?
17.   Are new mapkeys stored automatically? Where?
18.   What is the purpose of a part template? Where are they stored and how do you access them?
19.   How can you prevent the Browser or Navigator pane from opening automatically when you launch Creo?
20.   Are mapkeys case sensitive?

Exercises

1.   Create an assembly template. This should have named datums and named views to match your view selection mapkeys and default units to match your default part template. Make this the default template for assemblies.

Project Exercises

We're going to start off with some of the easier parts in the cart. These should give you some time to experiment with your configuration file, mapkeys, and part template. The project parts are shown in the figures below. Their location in the cart is also shown here for reference.

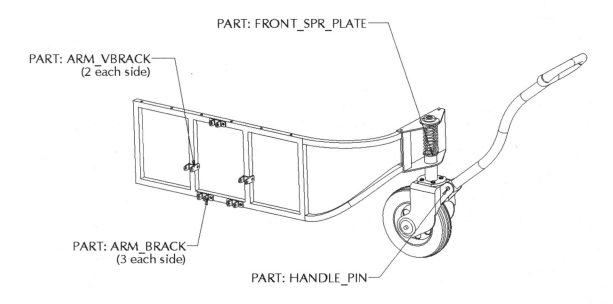

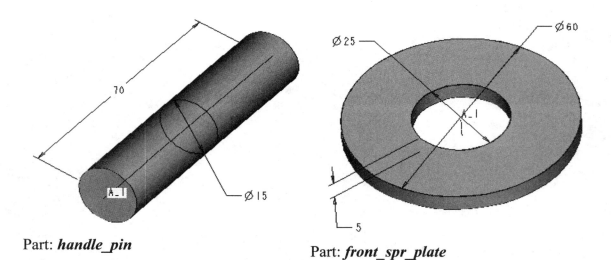

Part: *handle_pin*   Part: *front_spr_plate*

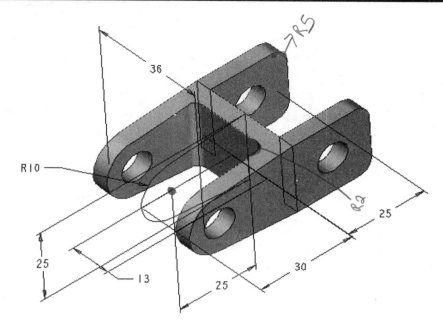

Part: *arm_vbrack*

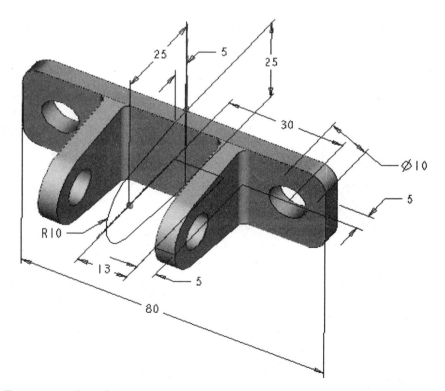

Part: *arm_brack*

This page left blank.

Lesson 2

Helical Sweeps and Variable Section Sweeps

Synopsis:

Helical sweeps; pitch graphs; variable section sweeps; using multiple trajectories; sweep parameter *trajpar*; using a *Datum Graph* to control a swept section

Overview

This lesson will introduce you to some of the main aspects of advanced sweeps, including helical sweeps and several different forms of variable section sweeps. Helical sweeps are quite special in that the trajectory, as you will have guessed, forms a helix. There are some interesting variations on this theme. Variable section sweeps are extremely flexible features, and offer quite a number of options: using multiple trajectories, relations, built-in parameters and functions (*trajpar* and *evalgraph*). We will also see how to create and use a *Graph* feature.

The main purpose of a sweep (and its primary difference from a blend) is to provide precise and known control over the cross section of the feature along its entire length. As you probably know, a blend is created by specifying known cross sections at various positions. Creo then joins these using either ruled or smooth surfaces. The shape of the blend between sections is not explicitly known or specified. Not so for a sweep, where you have precise control over the cross section at all locations. An example of this is the cam profile we will create at the end of this lesson. Sweeps tend to give smoother and more predictable variations in surface curvature, which is desirable when designing many consumer products. Sweeps are also more variable than blends in the orientation of the section at various locations along the trajectory.

The primary difficulty in modeling with variable section sweeps is identifying what is required, and choosing how to set up the references so that the sweep has the geometry you want. After some experience with them and more familiarity with the options, the job will get easier.

Helical Sweeps

A very common request of students in a mechanical design course is "How can I make a spring?" The answer is to use a helical sweep. This is a relatively straightforward feature that has the basic elements shown in Figure 1. These are the *profile*, an *axis*, which together define a revolved surface, and a *section* which moves up and around the surface along a helical path. If the profile is parallel to the axis, then a constant diameter helix is produced.

The profile is an open curve created on a sketching plane using Sketcher. The profile sketch also includes the axis of the helix. Revolving the profile curve 360° around the axis produces a surface of revolution that defines the envelope of the helical trajectory of the feature. The *pitch* of the helix is the distance traveled along the axis in one complete revolution of the helix around the axis. The number of turns of the helix is determined by the pitch and the projected height of the profile

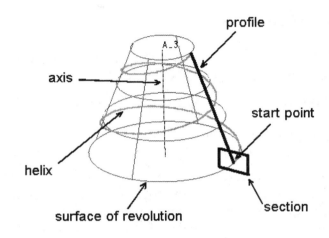

Figure 1 Basic elements of helical sweep

along the axis of the helix. In Figure 1, this worked out to exactly two complete turns. The section is a closed curve, also created using Sketcher, with the sketching plane located at the start point of the profile. Note that neither the surface of revolution nor the helix itself appear in the model as a surface or curve.

There are several main attributes that define the geometry:

- ▸ Solid or Surface
- ▸ Pitch: constant vs variable
- ▸ Section: orientation relative to the helix (***Through Axis*** or ***Normal to Trajectory***)
- ▸ Direction: right hand or left hand helix

Some examples of helical sweeps are shown in Figure 2. Although these are all protrusions, you can also use helical sweeps to create cuts, like threads on a bolt or power screw.

In this lesson, we will create two helical sweeps. The first is a simple one, using a constant radius profile and constant pitch, to illustrate the general procedure (similar to the sweep on the far left in Figure 2). The second will be a bit more complicated, using a multi-segment variable profile and variable pitch (a combination of the 2nd and 4th sweeps in Figure 2).

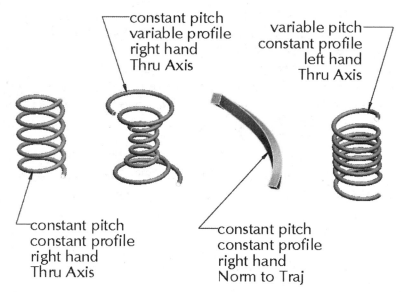

constant pitch
variable profile
right hand
Thru Axis

variable pitch
constant profile
left hand
Thru Axis

constant pitch
constant profile
right hand
Thru Axis

constant pitch
constant profile
right hand
Norm to Traj

Figure 2 Some examples of protrusions using helical sweeps

Constant Profile/Constant Pitch

Use the start part from the previous lesson to create a part called **hsweep1**. Then, in the **Model{Shapes}** group, **Sweep** overflow menu select

Helical Sweep

This opens the helical sweep dashboard shown in Figure 3. The default is a **Right** handed, solid, **Through Axis**, constant section, constant pitch sweep. Let's see what these options mean and where they are set.

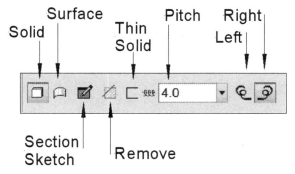

Solid Surface Thin Solid Pitch Right Left

Section Sketch Remove

Figure 3 The *Helical Sweep* dashboard

The *Solid*, *Surface*, *Thin Solid*, and *Remove* buttons behave as usual. The buttons *Right* and *Left* refer to the direction of the sweep, i.e. either a right- or left-handed helix. The *Section Sketch* button is used to launch Sketcher to create the cross-section that we want to sweep. The text entry area lets us specify the pitch of the helix (the distance advanced along the axis for each revolution of the helix).

The **References** tab is currently highlighted - we need to specify the helix profile. Open the tab. We can either select or define the profile, which we will do in a minute. At the bottom of this panel we can select either the *Through Axis* option (default) or *Normal to Trajectory*. This option determines the orientation of the section sketch relative to the helix. *Through Axis* means that the plane containing the section being swept will always intersect the axis of the helix as the section moves along the helical trajectory. The section, therefore, is not normal to the trajectory and the sketching plane for the section is oriented the same as the sketch plane for the profile. The other option (***Normal to Trajectory***) will result in the section being defined and maintained normal to the helix as it is swept along the trajectory. In this case, the section sketching plane and profile sketching plane will be different. Note that the difference between the resulting features will be more noticeable as the pitch to helix diameter ratio increases, as seen in the 3rd sweep of Figure 2.

The **Pitch** tab will eventually let us select between a constant pitch (default) helix, or specify the pitch points along the axis. The **Options** tab lets us cap the ends (if making a surface sweep) and specify whether the swept section is constant or variable. We will only use constant sections here.

As usual, most options can be selected in the RMB pop-up menu in the graphics window. Use that now to select ***Define Internal Helix Profile***.

We use Sketcher to create the profile. Use the **FRONT** datum as the sketching plane, and the **RIGHT** datum as the **Right** reference (this is the default). Create the sketch shown in Figure 4. Also, don't forget to create the axis (aligned with the RIGHT datum) using the RMB ***Axis of Revolution*** command.

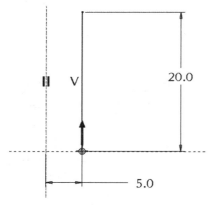

Figure 4 Profile for simple helical sweep

Figure 5 Finished constant pitch helical sweep

Note the direction arrow coming from the start point on the lower vertex of the sketch. If the arrow is at the top, select the lower vertex and select ***Start Point*** in the RMB pop-up. When the sketch is complete, accept it. A default pitch (probably 2.0) has been selected for you. Go to the pitch text box in the dashboard, or **Pitch** tab, or just double-click on the current value in the sketch, and enter **4.0**. How many turns will the helix be? This is the height of the profile (20mm) divided by the pitch (4mm) - exactly 5 in this case. Now use the RMB pop-up menu again and select ***Helix Cross Section***.

Since protrusions come towards you out of the screen, the view reorients so that you are actually looking at the back (dark side) of the FRONT datum. Crosshairs will appear at the start point of the trajectory.

Create a circular sketch (diameter **1.5**) centered on the crosshairs. Accept the Sketch and the feature will preview. The feature should look like Figure 5.

To explore the section orientation option, change the helix pitch to **10.0**. Go to the RIGHT view and zoom in on the start or end of the helix. The section sketching plane is aligned with the axis, as specified by the option ***Through Axis***. In the dashboard **References** tab, change the helix type to ***Normal to Trajectory***. You should see the end face of the helix is now perpendicular to the helix, and therefore no longer aligned with the axis. This is the type of helical spring you would get, for example, if you wound the spring using a round wire. The difference between the two options is very subtle unless the pitch of the helix is large. Change the pitch back to **4.0** using ***Through Axis***. Accept the feature and save the part. Investigate what happens when you use ***Edit Dimensions*** (in the mini toolbar) to change the spring dimensions and parameters. Where do these appear?

If you want to finish this spring off, you can trim the ends normal to the spring axis ("squared ends") as in a compression spring, or add hooks (perhaps using a variable section sweep covered later) for a tension spring. You might try to add relation(s) to automatically compute the pitch given the number of coils and the profile height. It wouldn't take much to set up some relations for the design of a helical coil spring with specified parameters like stiffness or allowed stress. We will discuss a module of Creo called Pro/PROGRAM a bit later that would assist with this.

Now we'll create something a bit more complicated...

Variable Profile/Variable Pitch

Start a new part called **hsweep2**. Select (in the **Sweep** overflow menu)

Helical Sweep

As before, use the RMB pop-up menu to create the helix profile and axis using **FRONT** as the sketching plane, and **RIGHT** as the **Right** reference. Sketch the profile shown in Figure 6. Note the start point at the bottom of the sketched profile and the profile is C1 continuous. Since we are using variable pitch, we will create a profile with a number of vertices. The pitch will be defined at several of these vertices. The minimum number of vertices is two, with

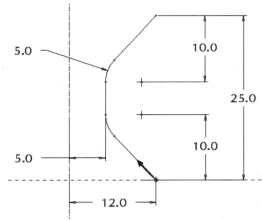

Figure 6 Profile for variable pitch helix (Sketcher constraints turned off)

(possibly) different pitch values at each end of the profile. Our sketch contains 6 vertices, and we will define the pitch at 4 of these locations.

When the profile sketch is complete (don't forget the axis!) and you leave Sketcher, you can adjust the pitch value at the start of the profile. Enter **5.0** then open the **Pitch** tab and click the button to *Add Pitch*. This will be for the other end and will also have a value of **5.0**. Click *Add Pitch* again; a marker will appear on the axis of revolution, with a location **By Value** from the start point. Instead of **By Value**, pick *By Reference* and select the vertex of the profile where the lower arc is tangent to the vertical segment. Set the value to **2.0**. Add another pitch point on the vertex at the top of the vertical edge of the profile. There are now four pitch points. Close the **Pitch** panel in the dashboard.

Open the **References** tab and select *Normal to Trajectory*.

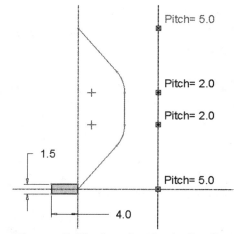

Figure 7 Section for the helical sweep

In the RMB pop-up (check out the available commands), select *Helix Cross Section*. The view reorients, and the crosshairs appear at the start point. If your datum planes are displayed, notice that your orientation is a bit "off." Recall that our section is going to be normal to the helix trajectory. Therefore, our sketch plane is at a slight angle to the default planes. Create a small rectangle whose vertical right edge is on the vertical crosshair as shown in Figure 7.

Accept the sketch and the feature should look like Figure 8. The profile and variable pitch are easy to observe in the FRONT saved view. Switch between *Through Axis* and *Normal to Trajectory* to see the difference in shape - not much for these pitch values. Accept the feature.

Check out how the feature appears in the model tree - the **Profile Section** and **Sweep Section** sketches are shown separately. This gives you easy access to attributes of the feature.

To change the pitch values, select the feature and then in the mini toolbar select *Edit Dimensions*. Change the pitch at each end to *10*. Now pick the profile curve and change the overall height to *30*.

Save the part and remove it from your session.

Figure 8 Variable pitch sweep

We've now seen most of the options involved in helical sweeps. Let's move on to something a bit more complicated!

Variable Section Sweeps

The idea for variable section sweeps is basically the same as simple sweeps: specify a trajectory and section. Simple sweeps are a special case of variable section sweeps (requiring, among other things, a constant swept section). The main differences of variable section sweeps are that the section can change size and/or shape along the trajectory, and the orientation of the section can change along the trajectory. Both the trajectory and section can be open or closed; if the section is open you can create either a surface or a thin solid. Variable section sweeps are among the most complicated Creo features. The main obstacle to using them is, perhaps, in visualizing how to apply the many options in order to create your desired geometry. Remember that sweeps can be used to create protrusions, surfaces, or cuts. The default is a solid protrusion.

In the following we will create several different parts using some of the different variations possible. The intent here is for breadth of coverage rather than extreme depth, and for you to understand the basic sweep forms and definitions of terms. We will see how to create simpler sweeps with constant or variable cross section, how to use multiple sweep trajectories as references, how to modify the section using relations, and how to use a new feature (a *Graph*) to provide data to control the section shape.

Some Definitions

It will help to understand the geometry of variable section sweeps if you clearly understand the following terms and how the geometry of the sweep is specified. See Figure 9. Refer back to these definitions as you do the following exercises.

Origin Trajectory - All sweeps must have one (and only one) of these. It basically provides the "backbone" of the sweep. The origin trajectory can be constructed from datum curves or feature edges. For variable section sweeps, this trajectory must be C1 continuous (all tangent edges with no kinks).

Frame - This is a Cartesian coordinate system whose origin moves along the origin trajectory. You will never actually see this system, but you must be aware of its orientation and how that is determined along the sweep. Its orientation is determined by its X- and Z- axes. The Y-axis can be determined using the right hand rule. The initial position of the frame is at the *start point* of the origin trajectory.

Section - A shape that is sketched on the XY plane of the frame. The sketch does not have to actually touch the origin trajectory anywhere. As the frame moves along the origin trajectory, the sketched section sweeps out a geometric surface. By capping the ends of this hollow surface, Creo can make a solid.

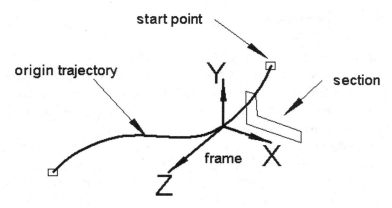

Figure 9 Main aspects of variable section sweep geometry

The next two terms refer to the frame's Z-axis:

Normal to Trajectory - if the frame's Z-axis is always tangent to the origin trajectory, then the swept section in the XY plane will be normal to the trajectory at the origin of the frame. This is the default. It is also possible to use other trajectories to define the normal direction. Only one normal trajectory is required, or allowed.

Normal to Direction - if the frame's Z-axis always points in a specified direction, then it will not, in general, be tangent to the origin trajectory. The swept section will always be normal to the specified direction. The direction is determined by the normal to a planar surface (making the swept section parallel to that surface), an edge, or two points. The specified direction for the Z-axis can not be perpendicular to the origin trajectory anywhere. (Why?)

Finally, a last term to know:

X-Trajectory - The orientation of the frame (i.e. rotation around its Z-axis) as it moves along the origin trajectory can be determined by specifying another trajectory which will be intersected by the X-axis direction of the frame. This is the X-Trajectory and is optional.

These are the essential elements of the variable section sweep geometry. We will look at several examples to try to get these sorted out and see how they work together.

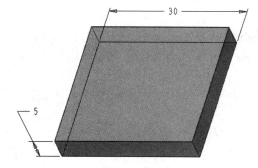

Setting the *Normal* Direction

The normal direction of a sweep is the easiest property to visualize. Create a new part called **vsweep1**. Create a 30 X 30 X 5 thick solid protrusion as shown in Figure 10.

Figure 10 Base feature for sweep

We are going to create the sweeps shown in Figure 11. Both sweeps use the same origin trajectory shape (a circular arc) and the same section shape (a 5 unit square). However, note the different orientations of the surfaces at the end of each sweep. These are determined by the normal direction for the section (Z-axis of the frame). In one case (on the left), the normal direction changes along the sweep; in the other case (on the right) the normal direction is constant.

These are quite simple sweeps, and we must supply only two elements: the sweep trajectory and a reference to define the normal direction. Other aspects of the sweeps are determined using defaults. Both these trajectories are planar, but general 3D curves are allowed (as long as the section does not become tangent to the trajectory!).

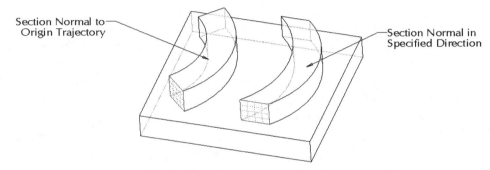

Figure 11 Two ways of specifying the normal direction. The swept section is shown with the grid lines.

Create a sketched curve on the top of the block, using the *Sketch* command in the **Datum** group. Use the top of the block as sketching plane, and the right face (or RIGHT datum) as the RIGHT sketching reference. Sketch the arc shown in Figure 12, making sure that only one horizontal dimension connects this sketch to the block. Accept the sketched curve. Create a copy of this using *Copy* and *Paste Special*, changing the 7 dimension to 22. The second datum curve is shown in Figure 13. Save the part.

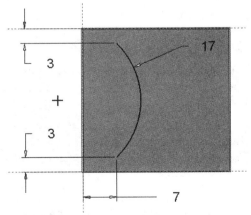

Figure 12 Datum curve for first sweep

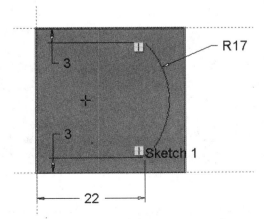

Figure 13 Copied datum curve for second sweep

Now we can get on with creating the first sweep (on the left in Figure 11).

We will use object/action command mode here: select the first datum curve (on the left), then pick the *Sweep* button in the **Shapes** group or mini toolbar.

On the model, the selected curve highlights in bold and is labeled as the **Origin** trajectory. An arrow at one end indicates the start point for the sweep (and therefore the sweep direction). Clicking directly on the arrow will switch it to the other end of the trajectory; if you do that, bring it back to the front. The start point is the initial position of the sweep frame (which is not shown) and is therefore the default location where we will sketch the section to be swept. Two additional labels (default 0) at each end can be used to extend the trajectory past the end of the designated datum curve. We will not use those controls in this lesson[1].

The sweep dashboard is also displayed. This is very similar to the helical sweep dashboard. The first two buttons on the left are toggles for creating a *Solid* feature (the default) or a *Surface*. The third button will launch Sketcher to create the sweep section. The next is a toggle for creating a cut, followed by a button to create a thin feature. The last two buttons are toggles for creating constant section (default) or variable section sweeps. Leave this set to *Constant* for now. Open the **References** slide-up panel. This will list all the trajectories used in the sweep and how they are used. In this example, the origin trajectory is listed, and the **N** option checkbox is selected. Thus, the default sweep is a *Normal to (Origin) Trajectory* sweep. Other details about the references and sweep options are given below the trajectory list box (**Section Plane Control**, and so on). In the **Options** panel, we see the two options (*Cap* and *Merge*) for treatment at the ends of the sweep. Leave these unchecked. The **Sketch Placement Point** is at the start point by default; if you have a datum point anywhere along the origin trajectory, you can place the Sketch Placement Point there (perhaps to take advantage of sketch references).

Since our origin trajectory lies on a plane, Creo can deduce enough information about the geometry so that we can immediately create the sweep section - no other section orientation information is required to define the sweep frame. As we will see, if the origin trajectory is not planar, additional information may be required to define the frame orientation (a default direction will be chosen for you).

Select the *Sketch* button in the dashboard toolbar (or in the RMB pop-up menu). The model rotates and we are looking at the crosshairs at the start of the sweep - these define the X (to the right) and Y (to the top of screen) directions of the frame. For a solid protrusion, the sweep will come towards you out of the screen. Sometimes the orientation of the model is a bit strange when you are first shown the section sketching plane and the crosshair. Spin the model to orient yourself and observe where the sketching plane is relative to the origin trajectory.

For this sweep, we will sketch a square with a vertex on the origin of the crosshair (i.e. touching the trajectory). This is not mandatory - the entire sketch can be offset from the trajectory if desired. Create the sketch (a 5 unit square) shown in Figure 14 (some of the constraints are not displayed in the figure).

---

[1] See Lesson #11 in the Creo Tutorial.

With a completed sketch, leave Sketcher. The sweep is now previewed (in yellow if you are in shaded mode). Note that the initial and final sections are normal to the trajectory. We could also have created this shape using a revolved protrusion (but would have to locate the revolve axis somehow).

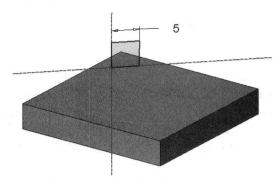

Figure 14 Section for first sweep

Before we accept this feature, try out the **Surface** and **Thin Feature** options on the dashboard. But, remember to select **Solid** before you finally accept the feature. The final accepted feature should look like Figure 15. If the feature shows in purple in wireframe (or light blue in shaded mode), you have created a surface instead of a solid.

We will create another sweep using the other datum curve, the same section shape, and a different normal direction. We will do things in the action/object order here.

With nothing highlighted, select the **Sweep** command. The dashboard opens. Click on the second datum curve; it becomes the origin trajectory and is highlighted in bold as before. Opening the **References** panel, the origin trajectory (default N checked) is listed. Open

Figure 15 First sweep completed

the pull-down list under **Section Plane Control**. The three options are listed. Read the tooltips for these options. Change to **Constant Normal Direction**. The panel content changes. The collector beside **Direction Reference** is active and waiting for us to select the normal direction reference. Pick on the front surface of the block (facing away from you in Figure 16) or the FRONT datum. The plane will highlight, and the normal direction arrow will appear.

We only need the origin trajectory for this feature; no other trajectories are necessary. Make sure the **Solid** option is selected, and launch Sketcher using the RMB **Sketch** command. Observe the orientation of the sketching plane relative to the block - it should be parallel to the designated **Direction Reference** plane (same normal direction). Create a 5 unit square section. The sketch and crosshairs are shown in Figure 16.

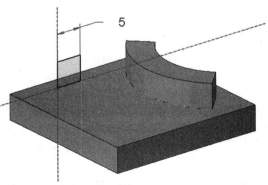

Figure 16 Section for second sweep

When you are finished with Sketcher, select *OK*. The sweep is previewed. The geometry is shown in Figure 17. The section does not rotate as it is being swept (as in the first sweep), but maintains a constant orientation. Note the difference in the appearance at the ends of the sweeps. We are going to play around with this feature a bit, so accept the feature and save the part.

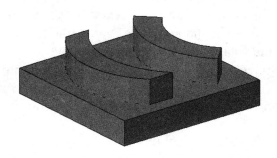

Figure 17 Second sweep completed

Use *Edit Definition* with either of our two sweeps and experiment with some of the options. For example, for the second sweep try changing the **Direction Reference** that defines the section normal. You can use other planar surfaces, or even edges, to define the normal direction. Some of the resulting sweeps may surprise you (try to figure out the geometry!) or may not be possible (in which case the sweep will not preview). For example, for the second sweep, try to set the **Direction Reference** to the front top edge of the base block. Why does this fail?

Cancel out of the dashboard. Before you leave this part, just for practice delete the two sweeps and create them again. Pay attention to the command options available with the mini toolbar and RMB pop-up menu - you may find you do not need to go near the dashboard at all.

Now, we will explore something a bit more complicated - using a second trajectory (the *X-Trajectory*) to control the orientation of the frame as it moves along the origin trajectory.

Using the *X-Trajectory* Option

The **X-Trajectory** option is used to control the direction of the frame X-axis as it is being swept. It allows the frame to rotate around its Z-axis (whose direction is determined using the Normal options given in the previous exercise). This is where variable section sweeps start to deviate significantly from simple sweeps.

Start a new part called **vsweep2**. The part we are going to create is shown in Figure 18. The basic sweep geometry is along a straight line. Several tear-drop shaped cross sections along the sweep are shown. Observe that these all have the same shape, but their orientation changes. To construct this sweep, we can use a straight curve for the origin trajectory (this is not mandatory) but must do something special to produce this rotation of the section. This is where the X-trajectory comes in.

Create two curves as follows (see Figure 19):

Curve #1 - Using the *Sketch* command, this curve is sketched 200 units long on the intersection of TOP and RIGHT. This will become the origin trajectory.

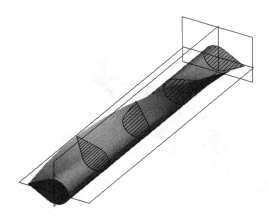

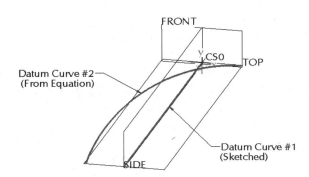

Figure 18 Sweep using X-trajectory to change section orientation

Figure 19 Two datum curves to construct sweep

Curve #2 - In the **Datum** group overflow menu, select the **Curve** command, then **Curve From Equation**. Open the Reference panel and select the part default coordinate system. If that is not there, you can create a Cartesian coordinate system CS0 (see Figure 19) on-the-fly while creating this curve. Note the Z-axis of the coordinate system is along our origin trajectory direction. Select **Cylindrical** in the first drop-down list in the dashboard, then the **Equation** button. Enter the following parametric equations (this should produce a curved datum that spirals around our first datum curve for half a turn). In these equations, the parameter *t* goes from **0.0** at the start of the curve to **1.0** at the end (these values are set in the two boxes in the dashboard).

```
r = 25
theta = t * 180
z = t * 200
```

With the two datum curves created, we can proceed with the sweep. We will use object/action here. Pick the two curves - the first one you pick will become the origin trajectory, so be careful of order here. Use CTRL to pick the second one. Then select the **Sweep** command in the mini toolbar.

The two trajectories have labels Origin and Chain 1 (Figure 19). The origin trajectory also indicates the start point with an arrow. In the **References** panel, both trajectories are listed. The origin trajectory has the N checked by default. In the **Horizontal/Vertical** drop-down list, select **X-trajectory**. The Chain 1 trajectory becomes the X-trajectory. Now go to the dashboard toolbar (or RMB pop-up) and launch Sketcher.

When you enter Sketcher, notice the crosshairs on the start point of the origin trajectory. There is also a snap point (small cross) at the start of the second trajectory. The horizontal crosshair (X-axis of the sweep frame) goes through this snap point. As the section is swept, this X-axis must constantly change direction so that it always goes through the X-trajectory at all points along the sweep. Create the sketch shown in Figure 20 - the point of the teardrop will indicate the x-direction for the sketch. Notice that the sketch does not have to touch either the origin trajectory or the X-trajectory.

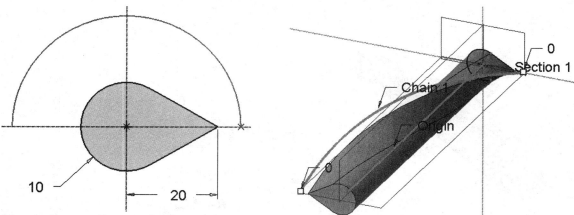

Figure 20 Section sketch for X-trajectory sweep

Figure 21 Sweep using X-trajectory

Accept the sketch. The sweep will be previewed as shown in Figure 21. Observe the point of the teardrop shape is always pointing at the second trajectory. Accept the feature and save the part.

To play with some of the options for this sweep, use *Edit Definition*. Remember that if you specify an illegal combination of options, all that will happen is that the preview will disappear. Also, note that to remove a trajectory, you must deselect any check boxes in the **References** panel associated with that trajectory first. Here are some things to try out (in the order presented):

*   Select the *Thin Feature* option in the dashboard. A slight change occurs in the dashboard that allows you to set the wall thickness and which side of the sketch (inside, outside, or both) the material is added. Turn this off again before proceeding.
*   In the **References** panel, deselect the X checkbox for the Chain 1 trajectory. This results in the sweep not changing orientation (becoming a simple linear protrusion). See Figure 22. Chain #1 really does nothing here.
*   Remove the definition of Chain 1. You cannot remove the origin trajectory, but you can replace it by selecting the second datum curve to be the origin. Use the first (straight) datum curve as the new X-trajectory. If the feature does not preview, open Sketcher and check the Sketch references. Note that this produces a *Norm to Trajectory* sweep.
*   Change the **Section Plane Control** to *Constant Normal Direction*, and pick the FRONT datum plane. Keep **Chain 1** designated as the X-trajectory. See Figure 23.
*   Return the trajectory definitions to our original ones, and edit the definition of the second datum curve to change the value 180 to 1800.

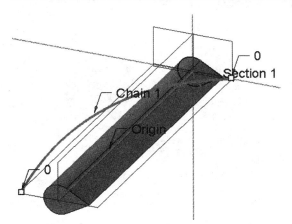

Figure 22 Sweep with X-trajectory turned off

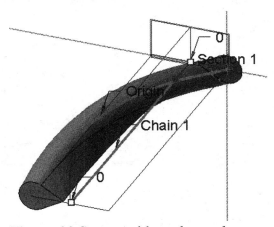

Figure 23 Sweep with exchanged origin and X-trajectory curves, Constant Normal Direction

Before we leave this feature and as a reminder, note the contents of the RMB pop-up menu. See Figure 23. This contains most of the common commands for setting sweep options. It is easy to get at these from here instead of opening the slide-up panels in the dashboard.

Save the part, and remove it from the session.

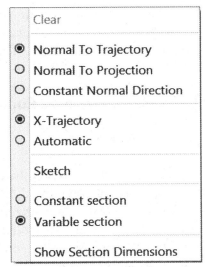

Figure 24 Short-cut pop-up panel using RMB

Auxiliary Trajectories and Variable Sections

Up to now, our swept section size and shape has been constant. We will now use additional auxiliary trajectories to control the size and shape of the section along the sweep. We will mostly use *Normal to (Origin) Trajectory* sweeps but are not restricted to that type. The section shape will change automatically in order to maintain some assigned dimensions or alignments to the auxiliary trajectories.

Start a new part **vsweep3**. For the base feature, make a ***Both Sides*** extruded protrusion off the RIGHT (or SIDE, depending on your part template) datum using the sketch shown in Figure 25. The feature depth is 80.

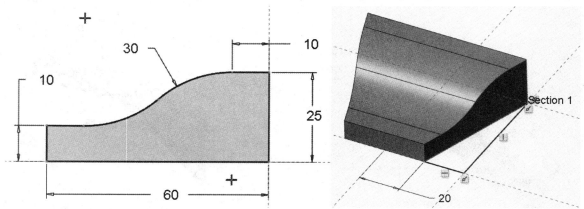

Figure 25 Sketch for base feature for *NormToOriginTraj* sweep

Figure 26 Creating cuts on end of base feature

Create a cut on one end of the base feature using the dimension shown in Figure 26. Create a mirror copy of this cut to the other end of the block.

Now we will create a datum curve to act as the origin trajectory. The curve should be at the intersection of the RIGHT datum plane and the top surface of the base feature. Using object/action, select the RIGHT datum, then select (in the **Editing** group)

Intersect

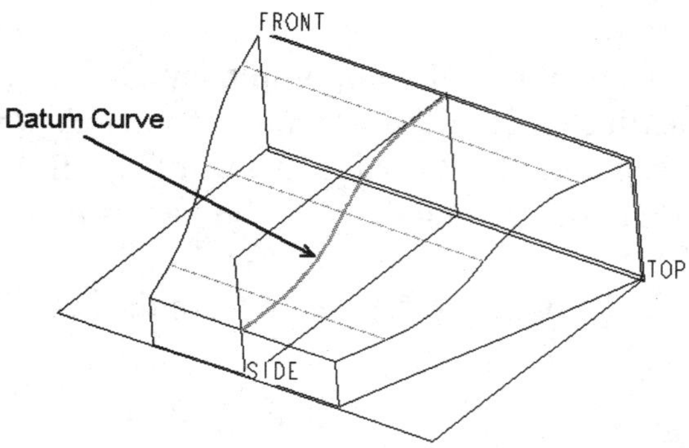

We are now in the **Surface Intersection** dashboard. Hold down the CTRL key, and pick on the four panels of the upper surface (easy to see in wireframe) to add each to the reference set. As each is picked, a yellow curve is created where it intersects RIGHT. When all four are selected, accept the feature. The curve should appear as shown in Figure 27. A single curve is added to the model tree (observe the feature name and icon).

Figure 27 Curve for origin trajectory defined using *Surface Intersection*

Now, create a sweep along this datum curve (see the final shape in Figure 31). Pick the curve, then launch the *Sweep* tool. If necessary, change the start point to the front of the block. Notice the default (**Normal to Trajectory**) listed in the RMB pop-up panel. We want to add an additional trajectory to the reference set. Hold down the CTRL key and click on just the front segment of the top-right edge of the base feature (there are four segments). It will highlight in bold, with the label Chain 1.

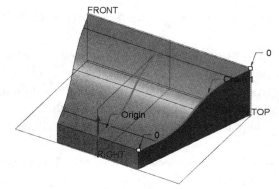

Figure 28 Auxiliary trajectory for sweep

Now, hold down (only) the SHIFT key and click on the segment at the back on the top-right edge. This will extend Chain 1 along the entire tangent edge (all 4 segments). See Figure 28.

In the dashboard, select the *Sketch* tool (or use the RMB pop-up). The model reorients so that we are looking at the crosshairs. You should be looking at the back of the block (remember that the sweep will come towards you) with the crosshair going through the start point of the origin trajectory. If the orientation is not clear to you, give the model a quick spin to see what you are looking at. Go back to the sketch with *View Sketch* in the Graphics toolbar. Observe that there is a snap point where the auxiliary trajectory, Chain 1, intersects the sketching plane.

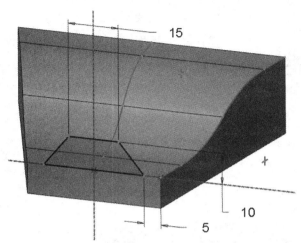

Figure 29 Sketch for swept section

Create a symmetrical sketch with dimensions as shown in Figure 29 (note that this is not the sketch view). Notice the dimension (5.0) to the snap point on the auxiliary trajectory.

Leave Sketcher and the feature preview should look like Figure 30. Observe that the default section orientation is Normal to Trajectory. Most importantly, the swept section does change shape as the side edges follow along the auxiliary trajectory at the specified distance.

Accept the feature and save the part. Come back into the feature with *Edit Definition* and play around with some of the feature creation options, such as

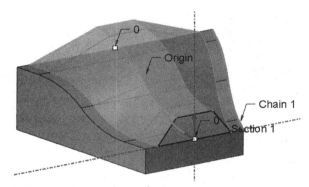

Figure 30 Previewed sweep

setting the Horizontal/Vertical control direction (the arrow at the start point). Either click on the arrow or use the *Next* button in the **References** panel. Study how these changes affect the final geometry. You may have to re-enter Sketcher to see what is happening and perhaps modify the sketch orientation. Before continuing, return the geometry to that shown in Figure 30 (or reload the part you saved previously).

Let's do a couple more sweeps to demonstrate some more ways to get variations in the section. We will create these along the end faces of this base feature.

For the first one, launch the **Sweep** tool and select the top right edge as the origin trajectory (use SHIFT to collect all four segments), as in Figure 31. Now use the RMB pop-up and select **Sketch.** Note that the sketch plane is not parallel to the front of the base. Why? Sketch a small closed triangular section with the apex on the origin trajectory and a height of 8. See Figure 32. The resulting sweep is shown in Figure 33. Note that the section shape is constant along the sweep. Accept the feature.

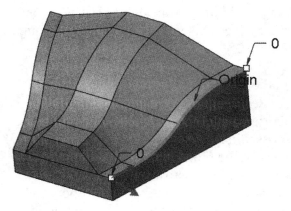

Figure 31 Origin trajectory

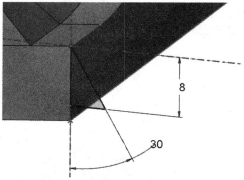

Figure 32 Section sketch

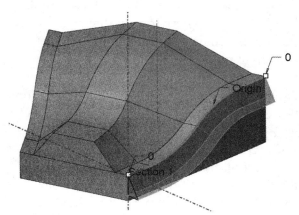

Figure 33 Sweep preview

Now we will add another edge to the trajectory list. Select the sweep, then in the RMB pop-up select **Edit Definition**. Open the **References** panel and click in the trajectory list box. Holding down the CTRL key, click on the lower right edge of the base feature. It will be labeled Chain 1. In the dashboard, select the button for a variable section. Now go into Sketcher. The auxiliary trajectory Chain 1 has created a snap point in the sketch. Redraw the closed triangle so that its base is aligned to the snap points at the origin and on the second trajectory[2]. Only the angle dimension should be required. See Figure 34.

---

[2] You can do this easily by setting an explicit alignment constraint to the snap point and deleting the 8.0 dimension when prompted to resolve the sketch. Make sure you pick the snap point and not the lower edge of the part.

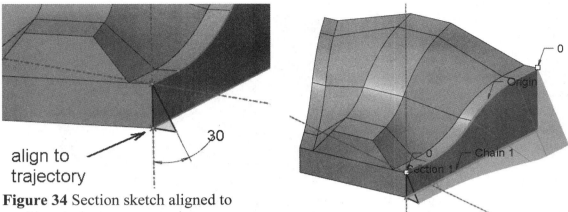

Figure 34 Section sketch aligned to auxiliary trajectory snap point

Figure 35 Sweep preview

Accept this redefined sketch and feature. The alignment is maintained as the section is swept. The angle of the sketch is maintained, and the width across the bottom grows to maintain the alignment.

Let's see what happens if we exchange the two trajectories. Create another sweep on the left face of the block. This time, use the bottom edge as the origin trajectory and the S-shaped edge as the auxiliary trajectory (Figure 36). The sketch must be closed, and keep the same angle dimension (30 degrees). Accept this feature.

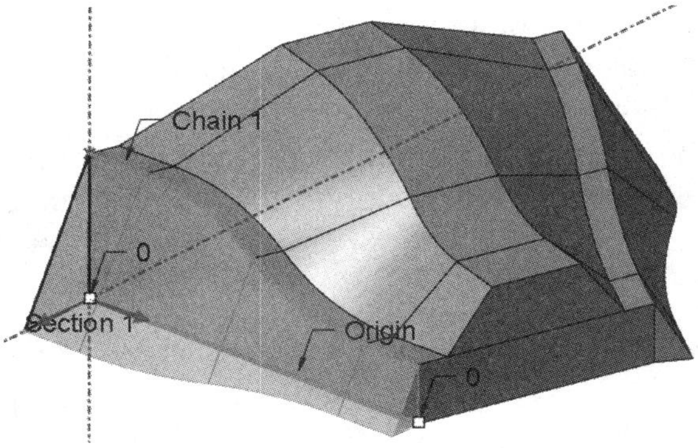

Go to the top view of the model (using the named views defined in the start part). A subtle difference exists between the sweeps

Figure 36 Sweep at other end of base

on the left and right ends of the block. It is easy to see if you go to no-hidden display with tangent edges turned on. What is the cause of this difference?

Highlight the last sweep and in the mini toolbar select *Edit Definition*. Then, again use the RMB pop-up and select *Sketch*. Explicitly dimension the horizontal width at the base of the triangle (this should make the angle dimension unnecessary). Change the width of the triangle at the base to **15**. The resulting geometry is shown in Figure 37. Note that the new lower left edge is straight. The angle of the section has automatically adjusted to follow the S-shaped edge of the auxiliary trajectory.

Accept the new sweep and save the part. See if you can modify the sweeps so that they are flush with the front and back surfaces of the base feature. This involves changing the **Normal** condition of the sweeps.

A couple of things you have to watch out for when creating variable section sweeps are to use a dimensioning scheme and values that will do the job that you want, and that will work for the entire length of the sweep. Try to keep your sweep section dimensions "local," i.e. within the sketch itself, and reference only the trajectories and section crosshair (that is, no other part geometry, surfaces, or edges). You will find out that the sweep will extend only as long as the shortest trajectory (i.e. as long as the references are all valid). This can be extended somewhat using the end

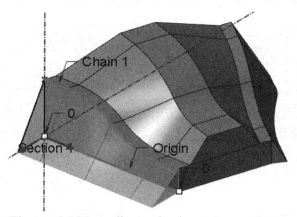

Figure 37 New dimensioning scheme for sweep

parameters (that we have left to the default value of 0 up to now) but only tangential to the trajectory at its end point.

As a final, and important note, remember that the *Normal To Trajectory* origin trajectory must be C1 continuous (all tangent edges). There cannot be any "kinks" in the trajectory where Creo will have trouble deciding on what the "normal" direction is. This is different from simple sweeps, where kinks in the trajectory can produce a mitered corner. Also, you can expect trouble if the section is large and straddles or is inside the curve of a trajectory with tight corners, where you run the risk of the sweep intersecting itself.

Using Additional Trajectories

Here is another example of including additional trajectories in the sweep definition. These "extra" trajectories can be used to control/define the cross section shape of the sweep along the main (origin) trajectory. Also, in this example, the auxiliary trajectories are not in the same plane.

Start a new part called **vsweep4**. Create three sketched curves as shown in Figure 37. The first curve is on the intersection of the RIGHT and TOP datums and is 100 units long. You can sketch this using appropriate datums as sketching and reference planes. Note that curve #2 (sketched on TOP) has equal length line segments at each end, with a circular arc in between. Curve #3 (sketched on RIGHT) is a simple arc.

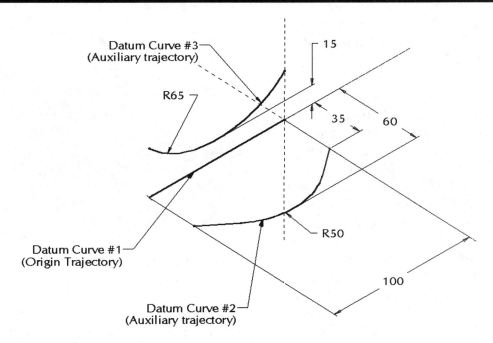

Figure 38 Datum curves

We'll create a sweep using these three curves to define how the swept section changes
shape along the sweep. Using object/action procedure, pick the three curves in order
(holding down CTRL). The first curve selected is automatically the origin trajectory. (See
Figure 39.) The other two curves will become auxiliary trajectories. Launch the *Sweep*
tool, hold down the RMB and select *Sketch*. We are now looking at the crosshair
(reorient the view to figure out where you are). Create the sketch shown in Figure 40,
aligning the three vertices of the sketch to the ends of the trajectories. Only one
dimension is required in the sketch for the arc. Select *OK* in Sketcher. The feature
previews as in Figure 41.

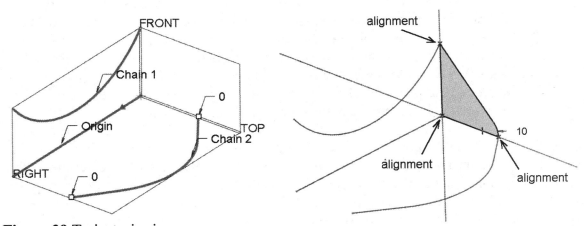

Figure 39 Trajectories in sweep   **Figure 40** Sketch of section

Looks sort of like Batman's hat! Well, maybe not...! Accept the feature and save the part.

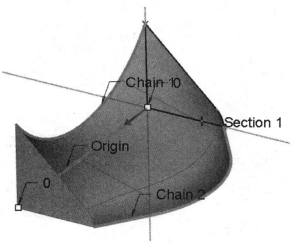

Figure 41 Sweep preview

Once the origin trajectory is determined, you can add as many additional trajectories as you want to be used in determining the section shape. You can expect that these are not totally arbitrary, since the resulting shape must still be a valid solid. For example, you cannot have a set of trajectories that would require the surfaces of the feature to pass through each other. There may also be geometries where, for example, the position or length of the auxiliary trajectories do not allow the determination of the section orientation along the entire origin trajectory. Try redefining curve #3 in the latest part so that it extends only from the FRONT datum to the bottom of the arc.

Variable Sections using *trajpar*

In this feature, we will use relations to drive the section geometry using a built-in parameter *trajpar*. This parameter is available for all variable section sweeps. Its value is 0 at the start point and varies from 0 to 1 along the origin trajectory (regardless of how many segments it has). The feature we are going to make is shown in Figure 43. This is one single solid feature! Using *trajpar* we can get many dimensions to change simultaneously in a coordinated way, without resorting to multiple trajectories.

Start a new part called **vsweep5**. We'll use a slightly different approach here to see what happens. Without any curves created yet, launch the *Sweep* tool. At the right end of the dashboard, select the *Datum* command, then *Sketch*. Create a single sketched curve similar to curve #1 of the previous exercise (100 units long on the intersection of RIGHT and TOP datums). When you accept the sketch, you are back in the sweep dashboard. Select the *Resume* button and the sketched curve is automatically chosen as the origin trajectory. The start point should be at the origin of the datum planes.

Use *Sketch* in the RMB pop-up to sketch and dimension the section shown in Figure 42. Note the symmetry constraints on the top vertices. To get these, create a centerline on the vertical crosshair and use the mirror command in the Sketcher toolbar (or use the explicit constraint). Make sure your sketch uses this dimensioning and constraint scheme. The dimension values are shown to the right of the figure.

While still in Sketcher, select (in the **Tools** ribbon)

> *Relations*

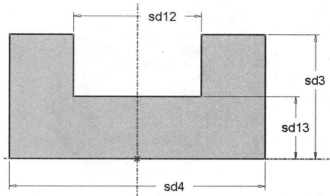

Figure 42 Section for sweep showing symbolic dimensions to use with *trajpar*

Dim	Value
sd3	30
sd4	60
sd12	30
sd13	15

NOTE: Your symbolic dimension names may be slightly different from those shown in the figure. Make a note of your symbolic names, as we will use these in some sketcher relations.

Enter the following relations (be sure to use your own symbolic names):

```
/* width of sketch
sd4 = 60 + 60*trajpar
/* height of sketch
sd3 = 30 - 20*trajpar
/* width of slot
sd12 = 55 - 25*cos(trajpar*180)
/* height of slot
sd13 = 10 + 5*cos(trajpar*180)
```

These will make the block grow wider and shorter along the trajectory, as well as modifying the width and depth of the slot. When these are entered, select **OK** in Sketcher; don't forget to select the **Variable Section** button on the dashboard. Accept the feature. The final shape is shown in Figure 43.

Where does the sketch for the origin trajectory curve appear on the model tree? We might have expected it to be inside the sweep feature, like a make datum, since we made it after launching the sweep tool. Curiously, that did not happen.

Save the part and close it out.

Figure 43 Completed sweep using *trajpar*

Using a *Graph* Feature to Control Section Dimensions

In the previous exercise we used explicit relations involving *trajpar* to control the section dimensions. This works well as long as the relation expressions don't get too complicated. For complex variations in a dimension along a sweep, it may be useful to use a *Graph* feature. In this exercise, we will use a *Datum Graph* to define a dimension on a cam to produce a desired variable offset of the follower as the cam rotates.

Start a new part called **vsweep6**. Create a
One Sided blind protrusion on the TOP
datum plane. Put a diameter 30 coaxial
hole in the center, as shown in Figure 44.

Now in the **Datum** group overflow, select

Graph

Enter a name *[camgraph]* for the new
feature. A new Sketcher window opens
where we will create the graph. First
create a coordinate system to identify the
origin of the graph. In the **Sketching**
group, select

Coordinate System

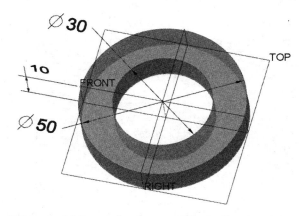

Figure 44 Base feature and hole for cam

and place a Csys towards the lower left corner. Create horizontal and vertical centerlines
through the origin of the coordinate system (these aren't mandatory but make
dimensioning the sketch easier). Then sketch and dimension the curve shown in Figure
45 (notice the left and right ends are horizontally aligned). The 360 dimension is critical,
as it represents one rotation of the cam. Note that all adjacent segments are tangent to
each other. You do not need to create the sketch exactly, other than the width being 360
and the curve being C1 continuous (all tangents between segments).

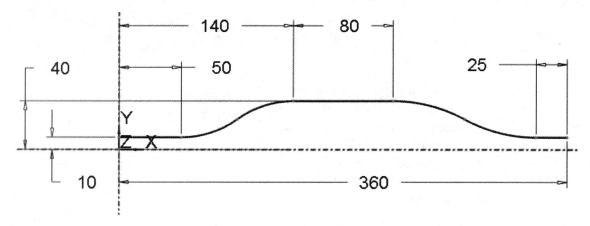

Figure 45 *Datum graph* to define cam follower displacement

In this graph, the horizontal scale (0 - 360) corresponds to the angular rotation of the cam
in degrees. The vertical distance will be used to define the outer surface of the cam,

resulting in a cam follower motion that will dwell-rise-dwell-return-dwell[3].

Select **OK** in the graph window Sketcher toolbar. Open the model tree to see the new graph feature.

Now we create the sweep using the graph. Launch the **Sweep** tool. For the origin trajectory, click on the bottom outer edge of the disk. Use SHIFT to get the entire circumference. No other trajectories are necessary.

Launch Sketcher and create a simple sketch which is a rectangle on the outside of the disk, aligned on one edge to the cross hairs, with a height of **8**. See the dimensioning scheme in Figure 46. The width of the rectangle (*sd4* in the figure) is controlled by the graph feature using a relation. Go to the **Tools** ribbon and select **Relations**. Enter the following (using your symbol for *sd4*) that uses a call to the built-in function *evalgraph*:

```
/* use evalgraph to determine dimension
sd4 = evalgraph("camgraph",trajpar*360)
```

The two arguments in the call to *evalgraph* are the graph name, and the horizontal position in the graph according to the graph scale. Note that since *trajpar* goes from 0 to 1, we have to scale up the second argument in the *evalgraph* call so that the graph is evaluated at the correct position.

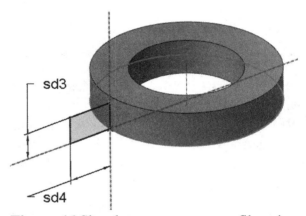

As an alternative, you could normalize the graph so that the range in both horizontal and vertical directions was 1. Then, in the above relation you could use *trajpar* directly since it represents a fraction of one revolution of the cam.

Figure 46 Sketch to create cam profile using dimension *sd4*

You could multiply on the right hand side by a scaling factor equal to the maximum real offset in units that match the part. The equation driving the section would then look something like:

```
/* use evalgraph to determine dimension
sd4 = 40 * evalgraph("camgraph_normal",trajpar)
```

If you made a library of graphs this way, you would have an easy way to create similar cams of arbitrary size all driven by the same graphs.

---

[3] This profile is not a particularly good one for a high performance cam due to the "jerk" (infinite derivative of acceleration) that occurs at the end of each arc segment. Better cam shapes result if the follower curve is constructed using higher order polynomials. This particular motion will also be produced only for a knife-edge follower. However, it will illustrate the construction method.

Select *OK* in sketcher, select the button to allow variable sections, and accept the feature. The final cam shape with the variable section sweep is shown in Figure 47. It is fairly easy, using the elements we have defined already, to create the cylindrical cam shown in Figure 48 by using the graph to control the other dimension in the section sketch instead.

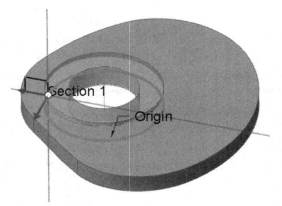

Figure 47 Preview of sweep on cam

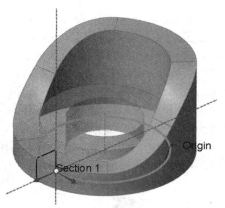

Figure 48 A cylindrical cam using the same displacement graph

Summary of Variable Section Sweeps

Variable section sweeps require, at a minimum, three elements: an origin trajectory, a section shape, and a way to determine the orientation of the section along the trajectory. We looked at two ways of specifying the orientation: using a normal direction, and using a horizontal reference trajectory (the X-trajectory). In addition, you can use additional trajectories to help define the variation of the section as it is swept. Other ways to do this involve sketcher relations, the trajectory parameter *trajpar*, and datum graphs. Clearly, there are many options here, complex geometry, and lots of flexibility. You will need lots of practice to be comfortable using variable section sweeps.

Conclusion

This lesson has introduced you to the main aspects of using advanced sweeps. This included helical sweeps and several different forms of variable section sweeps. The flexibility of variable section sweeps was illustrated using multiple trajectories, relations, *trajpar*, and *evalgraph* with a graph feature. There are numerous other options available. We did not touch on *Normal to Projection* sweeps at all, nor very complicated 3D trajectory shapes. You will want to experiment with these other options on your own.

In the next lesson we will look at the creation of advanced rounds, round sets and transitions, drafts, and ribs.

Questions for Review

1.   What are the essential elements in a helical sweep?
2.   What options are available to define the orientation of the section in helical sweeps?
3.   What are the restrictions on the section for a helical sweep?
4.   What are the restrictions on the profile for the two major kinds of helical sweeps?
5.   What is the pitch graph and where is it used?
6.   Can you create a helix that passes through or along the axis?
7.   What happens if the pitch is small enough that the sections of a helical sweep overlap? Can they touch, as in a close-wound spring?
8.   In a variable section sweep, what is the default orientation of the section relative to the origin trajectory?
9.   Explain the meaning and interpretation of the normal direction. How is it set?
10.   For the second sweep in the tutorial part **vsweep1**, if you use the left or right vertical face of the block to define the normal direction, the feature will fail. Why?
11.   What is the minimum number of trajectories that are required for (a) a default sweep, (b) a *Normal to Origin* sweep, and (c) a sweep of varying cross section?
12.   What is the maximum number of trajectories that can be used with a variable section sweep?
13.   What do you suppose will happen if any auxiliary trajectories intersect?
14.   How do you change the location of the origin trajectory start point?
15.   Edit the definition of the sweep in **vsweep2** to change the orientation of the section using *trajpar* in a relation (instead of using X-trajectory). HINT: this requires using a short construction line at an angle to the X-axis of the frame.
16.   How can you select a chain of solid edges to create a trajectory? Are there any restrictions on allowable edges?
17.   Can the X-trajectory have kinks?
18.   What is *trajpar* and what are its legal values?
19.   If a datum point is placed part way along the origin trajectory, it can be selected as the location for the section sketch. Since the sweep will go "both sides" from the section, what happens to the legal values of *trajpar*?
20.   How and where does a datum graph appear in the model tree?
21.   How is the shape of a datum graph interrogated in a relation (i.e. how is the graph evaluated?)
22.   Is it necessary to draw a datum graph at true scale in either, neither, or both horizontal and vertical directions? Explain.
23.   Within a single variable section sweep feature, can you combine the use of auxiliary trajectories, relations using *trajpar*, and a call to use a graph feature?
24.   Find out what a Möbius strip is, and see if you can make a model of this interesting shape. If you are really stuck for something to do, can you make a Klein bottle?
25.   How would you make the geometry shown in Figure 26 as a single feature? There are at least two ways using a variable section sweep.

Project Exercises

The spring that goes on the front wheel assembly is a simple, constant pitch helical sweep. The ends have been finished off with a revolved protrusion and a cut perpendicular to the spring axis. We need flat surfaces on each end of the spring to use in *Mate* constraints in the wheel assembly.

The two arms that support the side wheel mounting plates are shown below. Make the upper arm first using a simple sweep. Then do a *Save A Copy* to create the part file for the lower arm. You only need to edit the value of two dimensions to create the lower arm geometry. Using techniques we will discuss in Lesson 4, you could actually create both these, and many variations, from the same part file using something called a family table.

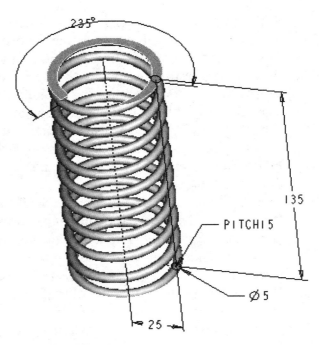

PART: *front_spring*

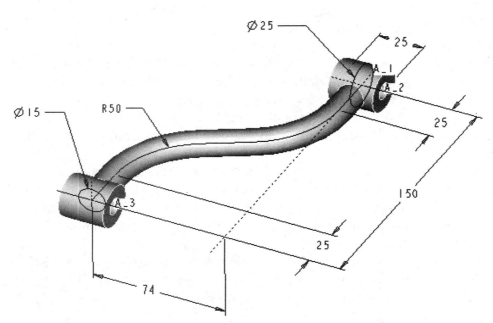

PART: *arm_upper*

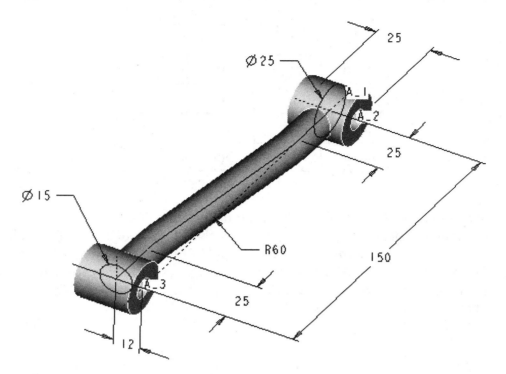

PART: *arm_lower*

The main square tubing frame members are based on parts created for the right side of the cart. The left side will be created in Lesson 8, using functions in assembly mode (*Merge* and *Mirror*).

First, create a default CSYS at the origin of the datum planes and dimension a set of four datum points according to the figure below. The datum point locations are indicated in the table on the next page.

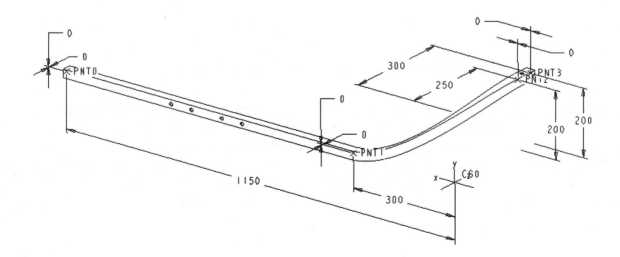

Point	X	Y	Z
PNT0	1150	0	0
PNT1	300	0	0
PNT2	0	200	250
PNT3	0	200	300

Create the datum curve for the lower tube using the datum points. Be careful to arrange the middle section of the curve to be tangent to the two end sections. HINT: start by making two separate straight curves, and do the middle curved section last. When you create it between PNT1 and PNT2 you can set the tangency option at each end. Project this datum curve (using *Project* in the **Editing** group) onto a *Make Datum* at a height of 350 above the XZ plane as indicated below.

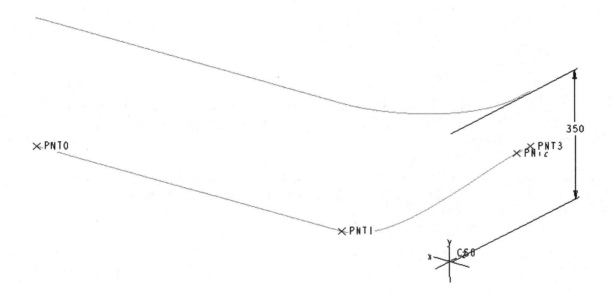

These two datum curves will form the origin trajectories of the two tubes.

Create the lower tube using a *Normal To Origin* trajectory. The lower three datum curves form the origin trajectory. Remember the procedure for picking tangent chains. The upper datum curve is the X-trajectory for the sweep. The sweep section is 25mm (outside) square, centered on the origin trajectory, with a wall thickness of 2.5mm. Use a *Thin* solid. The horizontal holes in the tube are created with a pattern table, which is discussed in a later lesson. You can put off making these holes until then.

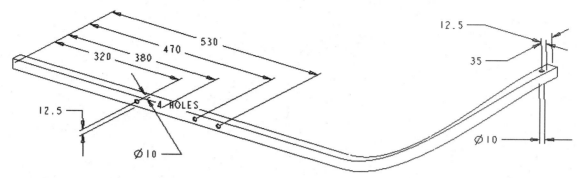

PART: *fram_low_rgt*

Finally, do a **Save A Copy** of the lower tube part to create a part file for the upper frame tube. In this new file, delete the lower sweep. Use the upper datum curve to create a simple sweep. The vertical holes in the tube are a simple pattern, starting at the left. You can create the horizontal holes for the bracket mount with a pattern table (do that later, too!).

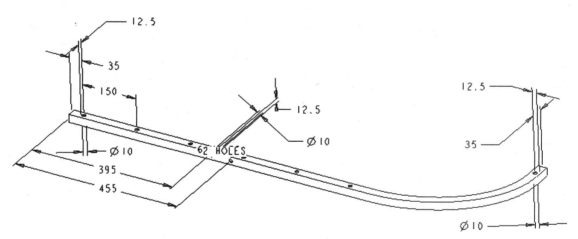

PART: *fram_upp_rgt*

This page left blank.

Lesson 3

Advanced Rounds, Drafts and Tweaks

Synopsis

Simple and advanced rounds (variable radius, thru curve, full round, round sets, round transitions); draft surfaces; tweaks (ribs, lips, and ears)

Overview

This lesson will introduce you to a number of "minor" features in Creo. The first half will be devoted to the subject of rounds. Rounds are often treated as cosmetic features and are usually added to the model towards the end of the regeneration sequence. Do not confuse "cosmetic" or "simple" with unnecessary and/or easy. We will see that the round functionality is actually quite powerful and can create very intricate geometry that is integral to the part function. The round command set is quite broad and deep. We will not be able to cover all variations here but should be able to show most of the commands for creating variations of simple rounds and how to set up and use advanced rounds with round sets and transitions. We will very briefly introduce the *Auto Round* tool.

In the second half of the lesson, we will create a simple part using several of the special solid creation commands formerly classed as "tweaks." The most important of these is for the creation of draft surfaces on parts destined for injection molding. Drafts can also become quite complicated so we will restrict our study to the basic principles. Towards the end of the lesson we will look at some very special purpose features (ribs, lips, and ears) that are referred to as *anatomic* features. Hopefully, with a better understanding of the terminology you will be better able to follow the on-line documentation.

Rounds

Rounds (or fillets) are usually among the last features added to a model. Most times they are purely cosmetic and are placed on the model simply to improve the visual appearance. However, rounds can sometimes be crucial to the geometry. We'll see an example of that in the second half of this lesson. For another example, a sharp concave corner of three edges is impossible to machine, so rounding these corners is more than cosmetic. People often have a tendency to add purely cosmetic rounds too early in the regeneration sequence just to make the model look good. Resist this temptation! Unless you have a good reason for creating rounds early, postpone round creation to the end. Purely cosmetic rounds are usually eliminated from the model (that is, suppressed) for purposes of finite element analysis - a process called "de-featuring." Adding rounds too early raises the possibility of setting up undesirable parent/child relations. Cosmetic rounds can also cause difficulties with view display in drawings.

A round feature consists of one or more *round sets*. Each set contains rounds sharing some common attributes (like radius). The default cross section shape is circular. This is often called the "rolling ball" shape obtained as a (hypothetical) ball rolls along in the direction of the edge staying tangent to the adjacent surfaces. There are several types of rounds based on how the round radius is determined. These are shown in Figure 1. You can also specify a conic shape in a couple of different ways.

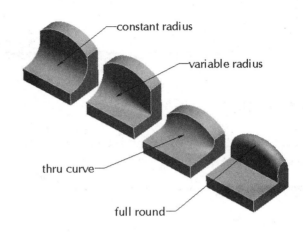

Figure 1 Basic round types ("rolling ball")

All the rounds within a single round feature are created simultaneously. When a round feature contains several round sets that interact with each other (at model corners), various kinds of *transitions* can be defined to specify the resulting geometry. The instant preview capability of Creo is very useful here to help you get the geometry you want.

It is doubtful that you will place all the rounds in a part in the same feature. So, you should still be aware that when rounds are created in separate features, the order of creation is very important in determining the resulting geometry. This is because the transition geometry depends on which round was created first.

Rounds are officially classified as placed features, requiring only a placement reference to determine their location on the model. Rounds can be placed using edges, surfaces, or some combination of these. Thus, the main job in creating rounds is to select placement references. Most other aspects are handled with reasonable defaults (you should know what these are and what other options are available). The graphics window preview will show the shape of the round prior to acceptance.

When editing or manipulating rounds, Creo will be in one of two functional and display modes:

Set Mode - this lets you create and manage the sets in the round feature. Within each set, you can place multiple individual rounds. Different sets can contain different types of rounds (constant or variable radius, through curve, full rounds, and so on). We will see examples of these in the first part of the lesson.

Transition Mode - when more than one round set is defined in a single feature and if the rounds meet each other, the geometry of the transition from one round to another can be specified using several options. It is also possible to trim a round at a specific point along an edge.

Three important parts of the interface for dealing with rounds are concerned with the use of *collectors*, selection methods, and RMB short-cut pop-up menus. Since a round feature can consist of multiple sets, each with multiple placement references, this is how you can keep everything sorted out. Usually, if you select something in a collector, it will highlight on the model (just like picking items in the model tree). You can also edit or remove items in the collector (using RMB pop-up). Selection is accomplished with a combination of left mouse button clicks and the SHIFT and CTRL keys as follows:

Left click	add selected entity to a *new* set
CTRL-left click	add selected entity to the *current* set
SHIFT-left click	add all tangent edges in a chain (or select a surface loop) containing the initial edge

Finally, you will find that almost all of the common commands for dealing with rounds are available in the RMB pop-up menu that appears in the graphics window. This includes shifting between set and transition mode, adding or removing sets, setting transitions, and so on. This menu is context sensitive. RMB short-cuts are also available in most of the collector areas.

There is a lot of variation possible with rounds, so we'll start slowly. This should get you comfortable with the interface commands and operation. Feel free to explore the various options as we proceed through each exercise. Start a new part called **rounds**, and create the simple block shown in Figure 2. We will do most of our exercises on the long, top front edge of this part.

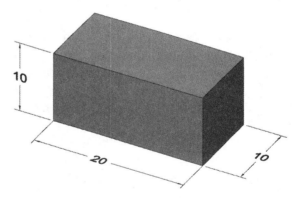

Figure 2 Basic block for round exercises

The On-Line Help

There is extensive on-line help for dealing with rounds. To access this easily, select

>   *File > Help > Creo Parametric Help*

then pick the links in the *Table of Contents* panel on the left as follows:

>   *Part Modeling > Part Modeling*
>   *Engineering Features > Round*

You can also use the global search function with the keyword "rounds" in the Part Modeling functional area to bring up a list of related help pages.

Creating a Single Round

In this section, we will look at variations possible for defining a round within a round set. It is possible to have several sets within the same round feature - we'll do that a bit later. The four round types we will look at are shown back in Figure 1: constant radius (including conic rounds), variable radius, through curve, and full round.

1) Constant Radius

The simplest round is constant radius along one or more specified edges. In the **Engineering** group select the ***Round*** button (don't confuse this with the chamfer tool which looks very similar; also note the shortcut key ***R***) and click on the top-front edge of the block. The edge will highlight in bold, and the round preview shows in yellow. Two "anchors" (small white squares) can be dragged to change the radius. Or, double click the indicated dimension and enter a new value, say **2.5**. The round dashboard is also open. The size of the round is also indicated there. To the left of the round radius are two buttons. These are toggles used to select either ***Set Mode*** (default) or ***Transition Mode***. Just below these buttons, open the **Sets** panel. It should appear as in Figure 3. This shows the sets collector (top left), attributes for the highlighted set, the references collector for the set, and the radius attributes below. A new round option is the ***Chordal*** round; try it out to see how the round is dimensioned.

The preview edges will be important later when we get to transitions in the advanced round section. To see what the actual round will look like, select the ***Verify*** button near the right end of the dashboard. Select this again (or middle click) to return to the round dashboard. The round feature is accepted by clicking the middle mouse button. As the last feature created, it will be highlighted.

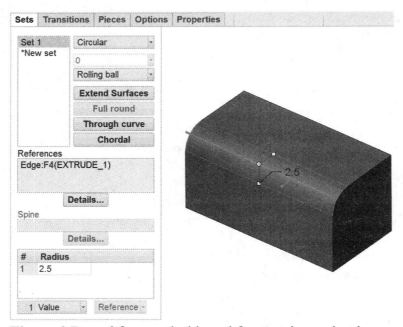

Figure 3 Round feature dashboard for creating a simple round

Now try the object/action method of adding rounds. Delete the existing round by highlighting it and picking **Delete** from the RMB pop-up menu. Then, select the top/front edge. Then either pick the ribbon button, or just pick **Round** from the mini toolbar. Now you can set the round radius and middle click again to accept the feature. How many operations were required to create the round? This is pretty quick.

The round set can also contain multiple edges. To change this round feature definition, select the round and in the mini toolbar, select

Edit Definition

This opens the round dashboard as before. In the **References** collector in the **Sets** panel, hold your pointer over the listed edge and in the RMB pop-up select **Remove**. Now left click on one of the four vertical corner edges in the graphics window. To add the remaining corner edges to the same round set (so that they all

Figure 4 Edited edges of first round

have the same radius), hold down the CTRL key while you left click on each edge. **If you do not hold the CTRL key down, you will create four separate round sets.** As they are selected, each edge will highlight in bold and the round will show in preview yellow. Observe that double-clicking on the radius value in the graphics window gives you access to a pull-down list of previously used radius values. Changing the radius value affects all rounds in the set. Set the radius to **2.5**. Accept the feature and the part should look like Figure 4.

You are probably aware of the importance of round creation order. To observe a selection behavior in Creo, we'll create a simple, constant round all around the top edge. Once again we'll use an object/action procedure. Select the straight top-front edge. It highlights in bold green. Now use the mini toolbar (or just **R**) to select **Round**. The entire tangent chain edge around the top surface is automatically added to the round set. Set the radius to **2.0** and use **Verify** to confirm the geometry. Have a look in the references collector to see how it is represented (only the first edge is listed). Since the first round feature modifies the corners, this tangent chain goes all around the top surface. See Figure 5. The key point here is that the default

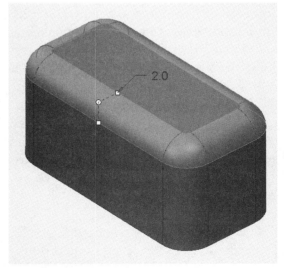

Figure 5 Round created on tangent chain around top of part

action of Creo is to extend the round placement reference along a tangent chain as far as it can (until it encounters a non-tangent corner). If you want to stop the round part way along the tangent edge (at an edge vertex or datum point, for example), you can use the **Pieces** options. Unfortunately, we don't have time for discussing that here.

To see some of the difficulties you might get into using simple rounds, change the radius of the top round to **4**. Notice the appearance of the round preview at the corners (Figure 6). Accept the feature. In the message window, you will get a warning about design intent - although the round can be created, there may be a problem. In the **Tools** ribbon select

Geometry Checks

and examine each of the listed items. The radius of the top round is so large that the round is self-intersecting as it goes around the corners. The advice given is to decrease the round radius. This is a common solution when you have round problems.

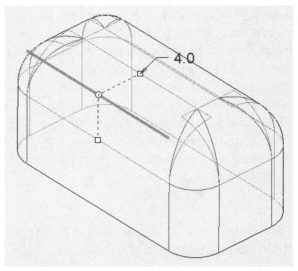

Figure 6 Top edge round radius modified. Note corner appearance

Observe (best seen in no-hidden wireframe) the geometry that Creo has used to go around the corner with the top edge round. Change the radius back to 2.0. This produces a different shape at the corner. These variations in corner geometry are called transitions, which we will deal with a bit later. Delete all the rounds.

To see some more aspects of selection, launch the ***Round*** command again. We will place a round on the four edges around the upper surface. Without the vertical edges being rounded these do not form a tangent chain. We will use a different short-cut to select them all. Pick on the top front edge. Recall that (1) if you pick on another edge with the left button, a new round set will be created and (2) if you hold down the CTRL key while left clicking on another edge, the new edge will be added to the current round set. We could use the latter method and pick on the four top edges. However, there is a better way! With the first edge selected and highlighted, hold down the SHIFT key and move the cursor over the top surface. A pop-up window will indicate the selection as a **Surface Loop** and all edges forming a loop around the surface will highlight. Clicking with the left button will now select all edges of the loop. Use the ***Verify*** button and observe the geometry at the corners where the rounds meet. This is an "intersection" type of transition, like a mitered corner. Have a look at this in both hidden line and shaded modes, then delete the round.

Before we continue, think about the three main topics we have covered so far. Understanding these clearly is important:
- launching the round command from the toolbar, mini toolbar, or keyboard shortcut, and using object/action procedures
- creation of round sets and layout of the **Sets** slide-up panel in the dashboard
- selection methods - manual and automatic and the use of CTRL and SHIFT

You should experiment with these aspects of the rounds interface before continuing. Another round attribute that you can investigate is the round geometry. This is changed by selecting, in the **Sets** slide-up panel, one of the conic options (see the shape pull-down list at the top of the panel). These are illustrated in Figures 7 and 8 below. The conic dimensions are the distance from the edge reference to the tangency point on the two legs of the round, and the conic parameter "rho" which determines the sharpness of the profile. Observe that these are listed in the **Sets** panel. The conic parameter must be between about 0.05 and 0.95; higher numbers produce sharper corners. For a perfect elliptical profile, the parameter rho must be 0.414... (enter this as "SQRT(2)-1"). If you haven't already, check out the ***Chordal*** option. Delete all the rounds before you move on to the next section.

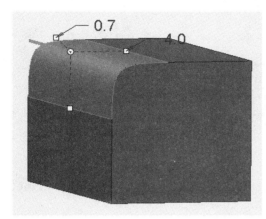

Figure 7 Conic round with D = 4.0 and rho = 0.70

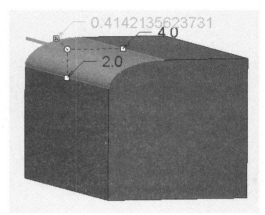

Figure 8 Conic round (D1 x D2 type) with D1 = 4.0, D2 = 2.0, and rho = SQRT(2)-1

2) Variable Radius Rounds

The previous exercises showed something about how individual edges or tangent chains and loops can be selected. The rounds in each set had a constant radius along their length. There are several ways to create rounds where the radius changes along the edge. Let's start with something simple. Select the ***Round*** command and pick on the top front edge of the block. The edge highlights in bold, and the constant radius value is shown in the dimension and in the radius collector in the **Sets** slide-up panel. A simple way to make the radius variable is to do either of the following:

- Place the cursor over any of the round anchors or dimension label. In the RMB pop-up, select ***Add Radius***, OR
- In the **Radius** collector in the Sets panel, in the RMB pop-up select ***Add Radius***

The current single radius value will replicate at both ends of the edge. The two new values are independently adjustable by dragging on the handles or entering a new value. Change them both to **4.0**.

We can also create additional radius control points along the edge. Place the cursor over any anchor, hold down the RMB, and once again select ***Add Radius***. This will add a new point somewhere along the edge. The two new dimensions give the radius value at the point, and locate the point along the edge using a fractional distance ranging from 0 to 1.

Both these values are adjustable by either dragging on the handles or entering new dimensions. Place new points at one-quarter and three-quarters of the way along the edge, and set the radius value to **2.0** at each point. See Figure 9. Open the **Sets** slide-up panel and observe the data listed in the collector at the bottom of the panel. To return to a constant radius round, select a cell in the first column, and select *Make Constant* in the RMB pop-up menu.

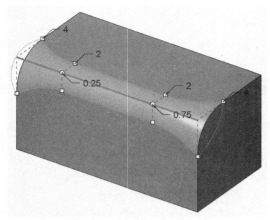

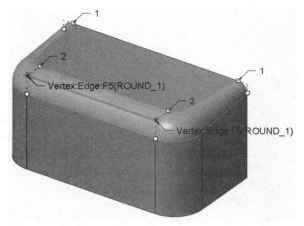

Figure 9 Creating a variable radius round

Figure 10 Variable radius round defined on a tangent chain

Variable radius rounds can be defined along tangent chains and edges. Each vertex at a tangency point can be selected to specify a round value and you can add more points if desired. Construct the model shown in Figure 10 by first creating the rounds on the two vertical edges at the front. Set their radius value to **3.0** and accept the feature. Then, create a tangent chain round along the top edge. Use *Add Radius* to convert it to a variable radius round. Set the end point radius values to **1.0**, then pick *Add Radius* again. The default action will be to add a vertex partway along one of the edge segments. You can drag the new point using the circular anchor that lies on the highlighted edge. To align the new point with a special vertex, open the **Sets** panel. Notice that an additional column (**Location**) has been added to the radius collector table near the bottom. All the radius control points for the round are listed here. Just below this table are two pull-down list boxes. If you select a row in the table (click in the entry in column 1), the list boxes show how the radius of that point is determined (by **Value** or by **Reference**) and where the point is located (by **Ratio** or by **Reference**). Change the location list box for the vertex we just added to **Reference**. You can now click on a vertex of the tangent chain to snap the radius control point to that location. Pick a vertex on one end of the front straight edge segment. Use *Add Radius* again to create another radius control point and align it to the other end of the same segment. Change both radius values to **2.0**. The model should look like Figure 10.

Let's look at another way to create a variable radius round. Delete all the rounds on the model.

3) Through Curve Rounds

If the round is going to have a complex geometry, it may be awkward to create it using a large number of points in a variable radius round as we did previously, plus we have no control over the radius between these points. There is a better way - using the *Through Curve* option.

On the top of the block, create a sketched curve as shown in Figure 11.

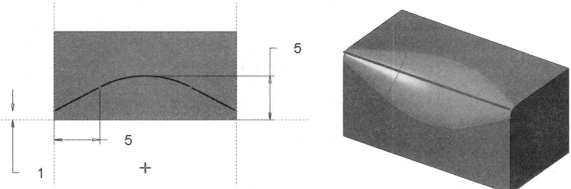

Figure 11 A *datum curve* to define round extent

Figure 12 A *Through Curve* round

We'll now create a round whose radius along the front edge is determined by the sketched curve. The new round will be tangent to the top surface at the location of the datum curve. We will use object/action procedure.

Select the top front edge, then select **Round** in the mini toolbar. Now in the **Sets** panel (or RMB pop-up), click the **Through Curve** button and pick on the sketched curve. It will be listed in the **Driving Curve** collector in the **Sets** panel. The round previews in yellow as usual (Figure 12).

The curve reference for a *Through Curve* round can be any curve, that is, including edges of the part. If a datum curve is used, it can be created on-the-fly during round creation (it will be grouped with the round, and hidden within the group).

Delete this round and the datum curve.

Yet another way to use datum features to drive round geometry is by using datum points. Create the four datum points shown in Figure 13. Select the top, front edge and **Round**. In the radius collector at the bottom of the **Sets** panel, select (in RMB pop-up) **Add Radius**. This puts radius values at each end and creates a variable radius round. Set the value at each end to **1.0**. Now select **Add Radius** twice more to create two more radius control points. Now, in the two pull-down list boxes below the radius value collector in the **Sets** panel, you can specify **Reference** for determining both the radius and the location of the new control points (listed as #3 and #4), then pick on the appropriate datum point. For example, the control point second from the left in Figure 14 uses PNT0 to determine the radius and PNT1 to determine the location.

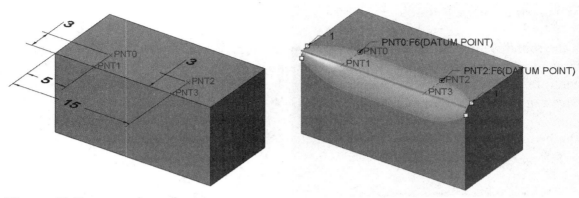

Figure 13 Datum point references

Figure 14 Round geometry driven by datum references

Delete the datum points (this will automatically wipe out the round).

4) Full Rounds

This is the last variety of simple rounds. The specification of a *full round* will result in a surface disappearing from the model. Full rounds are specified by picking pairs of edges or surfaces. If an edge pair is chosen, the surface containing both edges is removed. If two non-intersecting surfaces are chosen, it is necessary to specify which third surface joining the first two will be lost.

Let's investigate this by adding a cut to the base block to obtain the symmetric wedge shape shown in Figure 15.

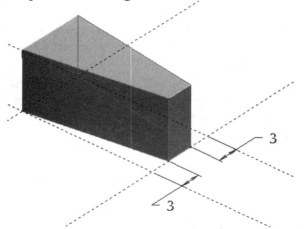

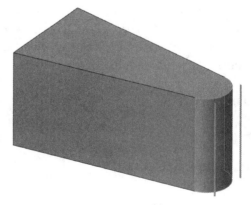

Figure 16 *Full Round* created on right end

Figure 15 Cut feature added to model

Pick the **Round** command. Select (using CTRL) the two vertical edges on the narrow end of the block. In the preview, observe the dimension value and that two rounds are created. Now, open the **Sets** panel and select **Full Round** (this is also available in the RMB pop-up menu). The resulting round is shown in Figure 16. The entire right surface has been cut off and replaced by a constant radius round going from the front to the back surface, and tangent to the (removed) end surface. No dimension is required to form this round.

In the graphics window, open the RMB pop-up and select *Clear*. This removes all the references from the current set. Now pick the two edges on the top of the part and the *Full Round* button again. This creates a variable radius round (Figure 17) and has removed the top surface. Once again, the round is tangent to the (removed) surface.

Both the foregoing are *Edge Pair* rounds; the surface containing the two edges is removed.

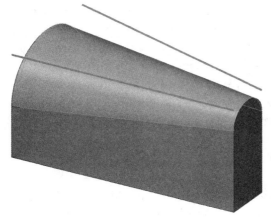

Figure 17 *Full Round* created on top surface

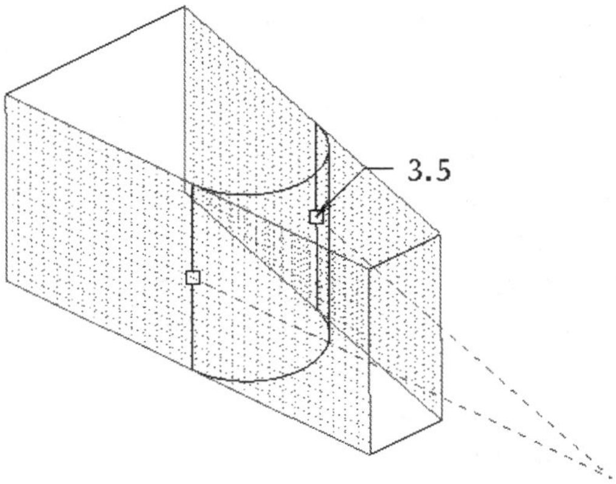

Figure 18 Creating a *Surf-Surf* round

Open the **Sets** panel and remove the references for the round (or use *Clear* in the RMB pop-up). Now select the front surface and using the CTRL key, select the back surface of the block. This forms a *Surf-Surf* round (Figure 18). Notice that a radius value is required. (What happens if you pick a radius too small? Try 1.0. What about a value of 2.2?) If you pick *Full Round* in the **Sets** panel, the radius will disappear and you are prompted to pick the surface that will be replaced by the round. There are four to choose from (top, bottom, left, and right sides - all of these share the two picked surfaces)! The one chosen will be listed in the **Driving Surface** collector. Accept the feature.

We have now seen the major round types available with simple rounds in a single round set. Some options have not been investigated here. For example, we have used *Edge Chain* rounds almost exclusively and have not touched on *Edge-Surf* rounds.

We have not encountered any seriously "diabolical" cases. Quite often, some trial and error will be required to get exactly the geometry you want. Creating rounds is something of an art, although it is becoming much easier with the capabilities of Creo. The more you know about the various alternatives for round creation, the better you will be able to create your desired geometry (without resorting to trial and error that is!). If you have trouble getting a round to create, the most probable cause is that the radius is too large. Try using a smaller radius and slowly increase it. You will probably be able to identify geometric obstacle to creating the round with the desired radius.

We will use the block in the next section, so delete any existing rounds, datums, and cuts.

Round Sets and Transitions

A round feature can contain a number of separate rounds organized into *round sets*. Each round set is defined using the commands introduced above. For example, in a single round feature you can have one round set containing several rounds with a constant radius, another set with a specific variable radius round, a third set containing a full round, and so on. This allows you to create several different types of rounds simultaneously, and have all rounds included in the same feature in the model tree. More importantly, because they are created simultaneously, the order that the round sets are identified in the feature does not matter. This removes a major headache when planning a round scheme for a part. We saw above that creation order is important for simple rounds in determining the geometry and appearance of the junction between separate rounds. When several round sets are defined at the same time, we have explicit control over the nature of the junctions between rounds, called *transitions*. Although default transitions will be used wherever rounds meet, we have numerous options to specify the transition shapes, which are independent of the round set creation order. We will look at some of these transition settings below.

We will demonstrate the main aspects of the use of round sets and transitions using our simple rectangular block. Delete all the other round features created in the previous section in this lesson.

Creating Multiple Round Sets

Creation of multiple round sets is largely a matter of using proper selection methods. To review (assuming we are picking on edges):

- left click - adds the chosen edge to a *new* round set
- CTRL-left click - adds the chosen edge to the *current* round set
- SHIFT-left click - finds a chain or surface loop containing the current edge

Recall that the default result of picking a segment of a tangent chain is to select the entire chain. In addition, you can always select *Add Set* from the RMB pop-up menu either in the graphics window or the sets collector. The **Sets** panel tells us which is the currently active set. The **References** collector shows what elements are contained in the current set.

We will create two round sets for our block. Launch the ***Round*** command.

Pick on the front right vertical edge and give a radius of **2**. This is all we will put in **Set 1**.

In the graphics window, use the RMB pop-up menu and select *Add Set*. Now (using the CTRL key) pick on the top front and top right edges of the block. Give them a radius of **2**. This is listed as **Set 2**.

Notice the default shape of the corner (Figures 19 and 20).

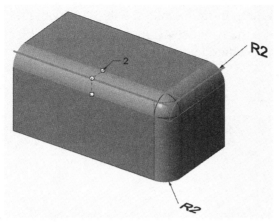

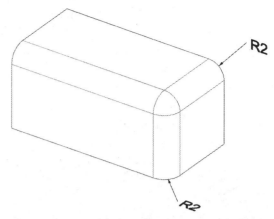

Figure 19 Multiple round set #1 (default transition)

Figure 20 Multiple round set #1 (default transition)

Select **Set 1** in the sets collector and change the radius of the vertical round to **3**.The corner transition has changed slightly. See Figures 21 and 22. This transition geometry is called a *corner sweep*. Different geometries (round sizes) will result in different default transitions. There are several available, and we will discuss transitions in further detail in the next section.

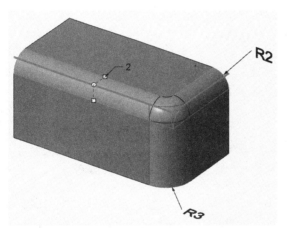

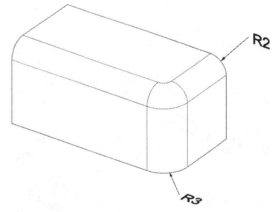

Figure 21 Unequal radii entering corner (transition is *corner sweep*)

Figure 22 Unequal radii entering corner (transition is *corner sweep*)

More Round Transitions

In the two cases above, we saw different corner geometries resulting from the default transitions. To investigate this further, on the **Round** dashboard select the second button from the left to put us in **Transition Mode** (or use *Show Transitions* in the RMB pop-up). The graphics display changes a bit. As you move the cursor across regions of the round, edges of several regions will highlight. The regions at the end of each round are called the stop regions. At the junction of the rounded edges is the transition region. Click on this region to select it.

Whenever rounds meet in a transition, a default shape will be used. This depends on the nature and size of the rounds entering the transition. Let's have a look at some of these.

Hold down the RMB in the graphics window. In the pop-up menu are listed the available transitions for this geometry. If you pick **_Intersect_**, the rounds will look like Figures 23 and 24. Each of the three rounds at the corner are extended until they intersect a surface of the other rounds.

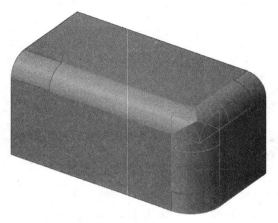

Figure 23 Intersection transition

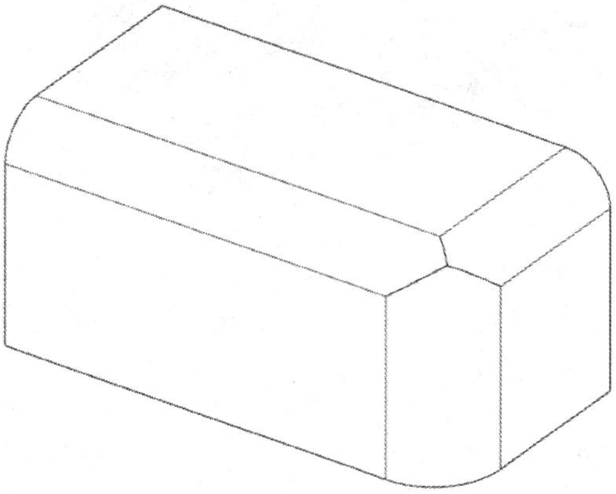

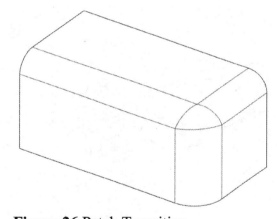

Figure 24 Intersection transition

In the graphics window, select the corner transition again, open the RMB pop-up menu and select the **_Patch_** transition. This is shown in Figures 25 and 26.

Figure 25 Patch transition

Figure 26 Patch Transition

Change the transition type to **_Corner Sphere_**. This complicated transition is composed of a number of pieces, as shown in Figure 27. The main component is a spherical surface whose radius by default is the radius of the largest round. Tangent to the sphere and each round coming into the corner is a blend. The length of the blend onto each round is governed by a dimension, as shown in Figure 28 (**L2** and **L3**, both set by default to the radius of the adjacent round, in this case, 2.0). These dimensions are also shown on the dashboard.

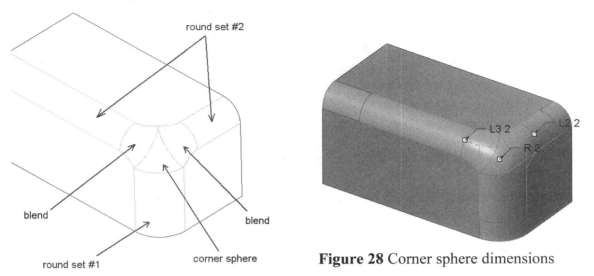

Figure 27 *Corner Sphere* transition

Figure 28 Corner sphere dimensions

With the corner sphere transition selected, return to **Sets** mode using the RMB pop-up and change the radius of the vertical edge to **2**, and the radius of the top edges to **3**. The top surface now has the larger radius rounds. The part appears as shown in Figures 29 and 30. Observe the change in shape and nature of the corner sphere transition. Experiment with different radius values and transition types. When you are finished, remove the part from session.

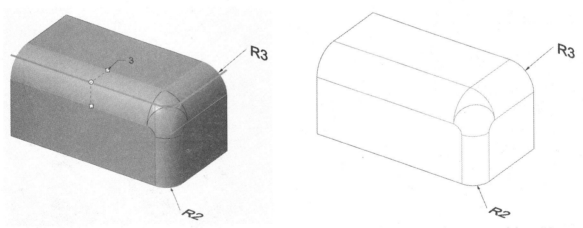

Figure 29 Corner sphere transition #2

Figure 30 Corner sphere transition #2

There are obviously a lot of possibilities for creating rounds. We have looked at a very simple geometry, with only two round sets defined on straight edges. These have all been convex edges. A round can also be formed on a concave edge (technically called a fillet). Creo's ability to produce rounds on difficult geometry is actually quite astounding. You are encouraged to consult the on-line help which has considerably more discussion. You will also learn a lot by creating test parts, such as those shown in Figure 31, and experimenting with combinations of parameters. For example, can you create the round geometry shown in Figure 32? This was done with a single round feature. What are the round and transition types?

Figure 31 A test part for rounds

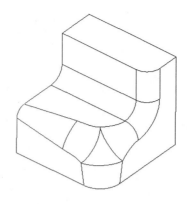

Figure 32 Rounds with transitions on test part

Another capability in Creo that you can experiment with is the behavior of rounds when the underlying geometry changes. This functionality is quite robust, with the rounds able to survive significant changes.

Another function in Creo you should check out is the ***Auto Round*** tool (see the ***Round*** command pull-down menu). This can be configured to locate and apply constant radius rounds to all concave and/or convex edges simultaneously. You have the option to select edges to exclude from this set - perhaps to create another Auto Round with a different radius. Check it out!

Drafts, Ribs and Tweaks

There are a number of special purpose features available in Creo that construct solid geometry for specific tasks. For example, if you are ever creating a part for injection molding, you will want to use the *draft* feature so that the part can be easily removed from the mold. *Ribs* are used to increase the strength of a part by creating thin reinforcing features between parallel or perpendicular solid surfaces. There are two versions of the rib feature (*trajectory* and *profile*) - we will have a look at each[1]. We will have a look at a couple of minor features, called *tweaks*. *Lips* and *ears* are very specialized features that could be made using standard protrusions and/or sweeps, but the construction demonstrated here is quite a bit quicker.

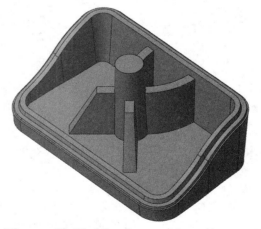

Figure 33 Finished part using ribs, drafts, and lips

---

[1] The trajectory rib feature can also contain draft, corner rounds where the rib meets the other part surfaces, and a full round on the top of the rib.

The part we are going to make is shown in Figure 33 which demonstrates several of these specialized features. This part actually contains only about a dozen features after the default datums. Although it is not immediately obvious from the figure, the part contains no vertical surfaces - these all have draft applied. The thin curved features between the cylindrical post and the back wall are trajectory ribs; the two with sloping surfaces on the front of the post are profile ribs. The treatment of the top sculpted edge is also easily accomplished using the lip feature.

Start a new part called **tweaks**. We'll proceed by creating a base feature, adding some rounded corners, shelling it out, and adding the circular post in the middle. In the first few steps of this exercise we will produce the geometry shown in Figure 35.

The first feature is a both sides blind protrusion off the RIGHT datum plane. The sketch is shown in the figure at the right and the feature depth is **24**. Make sure you select *Both Sides*.

Next, add some R3 rounds to each vertical corner edge. Note that although rounds are usually added last, in this case we will create them early so that the Shell command will also produce rounded corners on the inside of the part. In this case, the rounds are not simply cosmetic. Put all the rounds in the same feature.

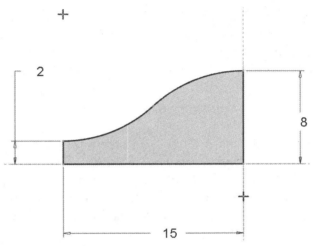

Figure 34 Sketch for base feature

Use the *Shell* command to remove the top surface, leaving a thickness of **1.5**.

Finally, create a one-sided solid protrusion of diameter **3** and height **10** coming up from the inside bottom surface in the middle of the part (align the sketch to the RIGHT datum). The part should now look like Figure 35. Now would be a good time to save it!

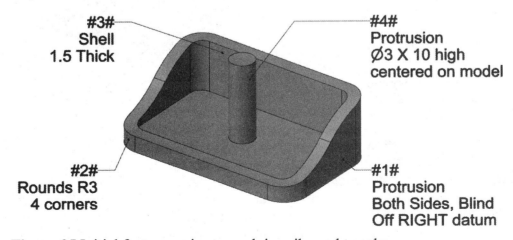

#3#
Shell
1.5 Thick

#4#
Protrusion
Ø3 X 10 high
centered on model

#2#
Rounds R3
4 corners

#1#
Protrusion
Both Sides, Blind
Off RIGHT datum

Figure 35 Initial features prior to applying ribs and tweaks

Ribs

As mentioned above, ribs are added to parts to increase their strength and stiffness. This is particularly true for thin-walled parts manufactured using plastic injection molding. There are two types of ribs in Creo: *trajectory* ribs and *profile* ribs. Trajectory ribs are similar to a one-sided thick protrusion that can be created using an open sketched curve of arbitrary shape and a Through Next depth spec. A profile rib (which is always straight) is also created using an open sketch but looking from the side of the rib. Our part will demonstrate both types of ribs.

Trajectory Ribs

We will first create the thin curved ribs between the back of the post and the shell wall. Since the rib is curved (looking down from the top), we must use a trajectory rib. To define the curved shape of the rib, we must create a sketch. We will do this on a make datum from within the rib dashboard. Use the *Rib* button pull-down and select:

> *Trajectory Rib*

In the dashboard select the **Placement** panel, then *Define* (or use the RMB pop-up command *Define Internal Sketch*). For the sketching plane, create a make datum that is offset a distance of **6.0** above the inside bottom surface of the part. Select the RIGHT datum as the right reference. Use the RMB pop-up menu *References* command and select the edge of the post and the inside back surface. Create the sketch shown in Figure 36. (Hint: This uses a short construction line inside the post profile to define the angle.)

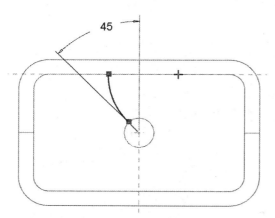

Figure 36 Single line sketch for trajectory of the first rib

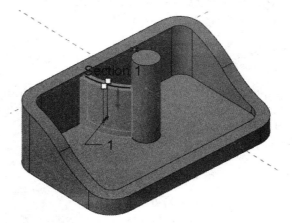

Figure 37 Preview of trajectory rib

When the sketch is accepted, enter a rib thickness of **1.0** using either the value showing in the graphics window, the dashboard, or the drag handles. This thickness is symmetric on each side of the sketched trajectory. If the rib does not preview, select the *Flip* button in the dashboard (or click on the creation direction arrow). To the right of the thickness value in the dashboard are three toggles that add draft, rounds on internal edges, and a full round on the exposed edge (top) of the rib. Check these out individually and in combinations. Open the **Shape** tab in the dashboard to watch the proceedings.

(IMPORTANT: Before accepting the feature, turn all these options **off**. Leaving them turned on will interfere with what we are going to do later with drafts.) Zoom in on the junction of the rib to the cylindrical post. You will see that the rib end has been merged to the curved surface.

Profile Ribs

Now we'll add a profile rib on the front of the post. As the name implies, the sketch shows the shape of the rib when looked at from the side (see Figure 38). We will again sketch this on a make datum. Open the **_Rib_** drop-down and select:

> **_Profile Rib_**

Launch Sketcher as usual to define an internal sketch and create a datum plane on-the-fly for the sketching plane. This should be through the axis of the post and at an angle (45°) to RIGHT. Pick the TOP datum as the top reference.

In Sketcher, specify the edge of the post, the top of the first rib, and the inner bottom surface of the shell as sketching references. Create the open sketch shown in Figure 38. When this is done, enter a rib thickness of **1.0** as before. Adjust the creation direction arrow so that the rib will preview. In the dashboard, note the 3-way toggle that will flip the rib to one side or the other of the sketching plane (check this out in the TOP view), or leave it symmetrically placed. Profile ribs do not have the same options for creating draft and rounds (this is offset by the ability to control the profile shape). Accept the completed rib and mirror both ribs through the RIGHT datum plane. See Figure 39. Notice that the rib has properly merged with the post, and the sloping surface is curved. The latter is because the rib has been merged to a cylindrical surface.

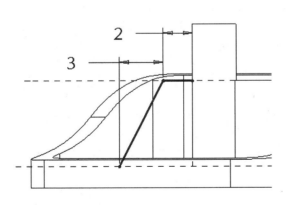

Figure 38 Sketch for profile rib

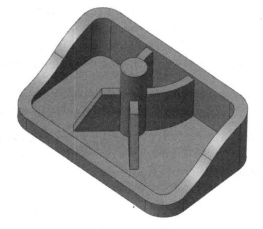

Figure 39 Ribs completed

Drafts

A draft is a modification to a surface so that the part can be easily removed from molds, as illustrated schematically in Figure 40. Without draft, the part will not leave the mold cleanly and/or may become damaged. Draft features can become quite complex. In addition to the part geometry, application of the draft requires knowledge of where the mold parting line on the part will be (where the mold halves come together), which may also be affected by the location of the gates where material is injected into the mold, and so on. Putting drafts on a part, unless they are fairly obvious, might be a task left to the molding specialist[2]. Consequently, in the following, we will look at only a few of the many options available.

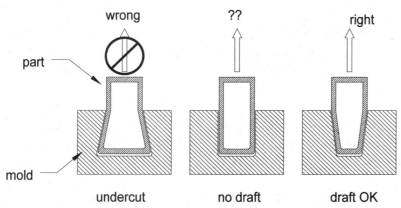

Figure 40 Draft surfaces on molded parts

Like rounds, drafts are placed features that are applied to existing surfaces. You can think of the draft as a small angular rotation of the surface about an axis or hinge line. By placing the hinge in the middle of the surface, you can create a *Split Draft* with different angles on each side of the hinge. You can also create a split draft by projecting a sketch onto the surface that divides the surface into two regions. A different hinge and draft angle can be specified for different regions of the surface. We will not discuss split drafts here. It is also possible to define a variable draft angle at different points along the pivot. Again, see the on-line documentation for more information.

An important point to note is that draft can be applied *only* to surfaces formed by planes or tabulated cylinders. If you need to put draft on a higher order surface, you may have to use a variable section sweep.

Like rounds, drafts are usually applied towards the end of the regeneration sequence. Drafts are normally applied before cosmetic rounds. In some cases, drafts applied to surfaces containing rounded edges will fail. So, if you need to round edges, the rounds should be applied after the draft.

To specify a draft, we basically need to provide four pieces of information (illustrated for a very simple draft in Figure 41):

• the surface(s) to receive the draft,

---

[2] On the other hand, some designers insist on adding draft themselves in order to be sure their design intent is not compromised. For example, draft causing a taper going the wrong way on a cylindrical part might cause an assembly interference or even functional failure. In this case, only the designer knows for sure!

- the draft hinge (the "Neutral Curve"),
- where the draft angle is measured from (the "pull direction" - normally the direction of mold opening), and
- the value of the draft angle (direction and size).

Some of these are interdependent. For example, if a plane is used to define the draft hinge (at the intersection of the draft surface and hinge plane), the pull direction will automatically be set normal to this plane (as in Figure 41). The pull direction is a bit ambiguous - it either represents the direction the mold will depart from the part, or the direction the part is ejected out of the mold. In either case, the draft angle can be used to pivot the surface in the correct direction, and confirmed by the preview or using **Verify**.

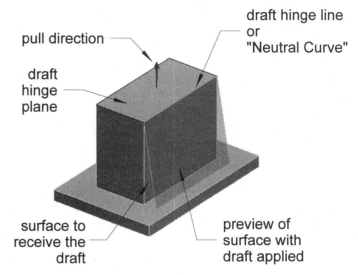

Figure 41 Draft terminology and geometry

Let's see how this works. We will use a variety of creation methods - both action/object and object/action, and explore some different selection methods and shortcuts.

First, we'll put some draft on the ribs. Although we can apply draft to each surface individually, all these vertical surfaces share the same hinge plane, so we will create them all at the same time. Select the ***Draft*** button in the **Engineering** group.

The draft dashboard opens. Read the message prompt. Holding down the CTRL key, select all the vertical faces of the ribs. There are eight surfaces in total. You can select these in any order. As each surface is picked it will highlight. Open the **References** panel to see the contents. Click the collector beside **Draft Hinges** (or the collector in the dashboard) and then select the horizontal top surface of one of the ribs. The intersection of this surface (plane extended) with each of the draft surfaces will define their hinge lines. The selected plane will also automatically define the pull direction as well (see the dashboard or References panel). The dashboard now contains a data entry field to specify the draft angle. The draft surfaces are now previewed. The pull direction can be reversed by clicking on the magenta arrow. The angle dimension of the draft (default 1°) is shown, with a drag handle that can be moved to adjust the draft angle or you can enter a new value (say 2°) in the dashboard. Note the ***Flip*** button. Set the draft so that the ribs are wider at the bottom than at the top.

TIP: Draft angles are typically very small, ranging from 0.5 to 2.0 degrees. These are hard to see on the model so can cause problems if you happen to make a mistake. When you first enter a draft angle, you might enter quite a large value, like 4.0, then

Verify the draft. If the direction is correct, change the magnitude of the angle to the desired value.

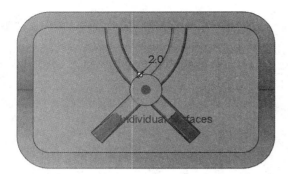

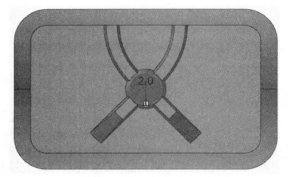

Figure 42 Top view showing draft on ribs

Figure 43 Top view showing draft on post

A good way to check the draft is to look down on it from the TOP view. This view is shown in Figure 42 which shows the edges at the top and bottom of the rib drafts. Before you leave, check out the other dashboard items. Accept the draft feature.

Now we'll add draft to the post using a slightly quicker way to select surfaces and use the object/action procedure. In the graphics window, select the front vertical curved surface of the central cylinder (only pick the front). Then pick the *Draft* command. In the dashboard, the **Draft Hinge** collector is active. Pick on the top surface of the cylinder. This automatically defines the pull direction and the draft surface preview is now visible. The draft extends all around the cylinder. This happens automatically when a chain of tangent surfaces is encountered (much like tangent chains when applying rounds). All you need to do is make sure the draft angle (2°) and direction are what you want, then middle click to accept the feature. The TOP view should look like Figure 43. Accept the feature.

We need some draft on the inner and outer vertical surfaces of the shell, keeping the same curved edges on the top (presuming that they must align with a mating part). The curved edges will become the draft hinges for these drafts. Since the edges do not lie in the same plane, we cannot use a plane to define the draft hinge (or pull direction), so we must do that a bit differently. Also, we must explicitly define the pull direction.

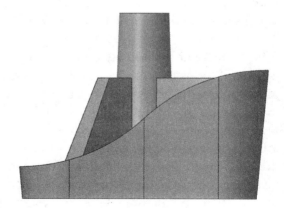

Figure 44 SIDE view showing draft on outer surface of base

Using the shortcut we observed on the cylindrical boss involving tangent surfaces, we can create the draft feature pretty quickly as follows. Select any single vertical surface on the outside of the base feature. We only need to pick one because all the rest form a tangent chain and will be added automatically. Now select the *Draft* command. The **Draft Hinge** collector is active. Pick on the top, back (outside) edge of the part. It highlights in bold. Now hold down the SHIFT key and pick on the other top outside segments. The entire tangent chain should now highlight in bold. This becomes the hinge line for the draft. Now, click in the **Pull Direction** collector in the dashboard, then select the TOP datum. The pull direction should be downward. The draft should now preview on all outside surfaces of the base. Adjust the angle to **5°** so that the base is narrower than the top, and accept the feature. The RIGHT view of the part should look like Figure 44.

Finally, we will put a 5° draft on the inner surface as well, using a variation of the surface selection method. In the lower right corner of the Creo window, change the **Filter** setting to *Geometry*. Now, right click on an inside vertical surface of the shell until the entire inside surface highlights and a tip window indicates "Intent Surface" (the shell surface). Left click to accept this, then select *Draft* in the ribbon. Specify the **Draft Hinge** using the same procedure we used for the draft on the outside (pick any inside top edge, then SHIFT-click to pick the other edges in the chain). Select the TOP datum as the **Pull Direction**. So far, nothing is previewed. The problem is that the horizontal inside surface of the shell is in our surfaces selection set. Open the **References** slide-up panel. Select the entry in the **Draft Surfaces** collector. Now CTRL-click on the horizontal surface. This excludes it from the surface set. The draft now appears in preview. Set the angle to 5° as before (so that the inner and outer vertical walls are parallel) and accept the feature. See Figure 45.

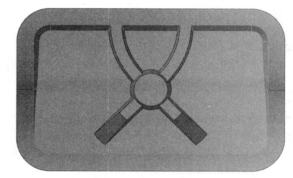

Figure 45 TOP view showing draft on inner surface of base

Here is something to think about that involves feature creation order. What would happen if we put the draft on the outer surface of the base feature before we shelled it and/or applied the corner rounds? What is the nature of the rounded surfaces on the corners of our part? How would this change if we created the draft first?

Since the trajectory rib can also contain draft within its own definition, you can expect that feature creation order may become important. This is particularly true if you also include automatically created rounds either where the rib joins the part walls or on its free surface. Some advance planning and perhaps some trial and error will be necessary here.

This has been a very brief introduction to drafts. As mentioned previously, the specification of draft features should perhaps be left to mold designers, who have the knowledge of best practices. In fact, mold makers often want to receive parts without any draft applied. This is not a problem usually, since drafts are among the last features applied to a part. Sometimes, however, it may be necessary (as it is for rounds) for reasons of design intent to apply draft much earlier.

Now, on to a couple of very specialized features. Save the part, but don't remove it.

Lips

IMPORTANT: To enable this and the next feature discussed, your configuration file *config.pro* must contain the following option:

  allow_anatomic_features   **yes**

A lip is a feature that is often added to plastic molded parts, such as an enclosure, where top and bottom parts must mate and lock together. The lip geometry is defined by the five items shown in Figure 46: the lip edge, the surface to be offset, the offset distance, the distance to the draft, and the draft angle.

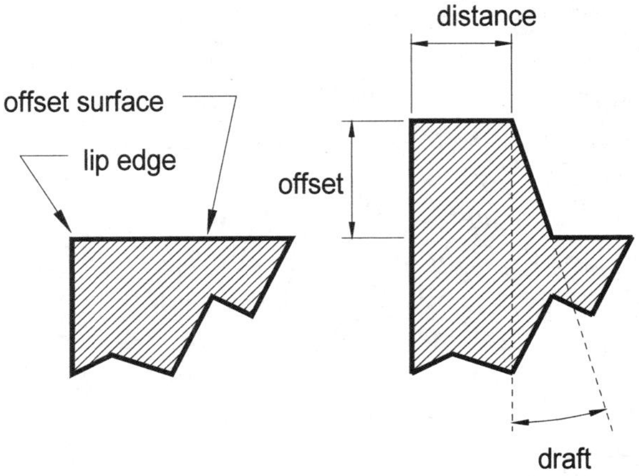

We will first create a lip around the inner top edge using a positive offset (as shown in Figure 46). Since the Lip command is not in the ribbon, go to the **Command Search** box at the top right and enter the text "Lip". Select the command, and we will enter some old-style (pre-Creo) menus at the right of the screen. Select **Chain**.

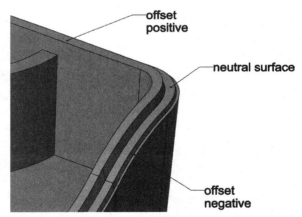

Figure 46 Defining a *lip* feature (left: initial edge geometry; right: edge with lip)

Pick on the top inner edge. The edge tangent chain will highlight in bold. Select **Done**. See the message window. Pick the surface to be offset (the neutral surface) as the horizontal top face of the wall at the back of the part. Enter an offset value of **0.5**. The distance to the draft surface is **0.75**. Select the TOP datum as the drafting reference (like defining the pull direction), and enter a draft angle of **15°**. The lip is now completed. Very simple!

Figure 47 Completed *lips* on top edges

Create another smaller lip on the outside edge that removes material using a negative offset[3]. Again use **Command Search** to select:

   Lip > Chain

Pick on the outer top edge. The surface to be offset is the same as before. The offset value is negative: **-0.25**. The distance to the draft surface is **0.25**. The draft reference is

---

[3] Putting this tiny bit of relief on an exterior edge is a trick sometimes used by mold makers to disguise the mold parting line.

the TOP datum and the draft angle is again **15°**. A close-up view of the two lips is shown in Figure 47.

The mating part would use similar lips but with the opposite sign offset on the larger inner lip to produce a mating connection between the two parts.

This completes all the features we want to add to this part. Save the part and remove it from the session.

Ears

This is another very specialized feature and is a really quick way of producing a thin angled tab on the edge of a part; see Figure 49. The sketch of the shape is done in the plane of a part surface and the angle of the tab is simply specified. Create a new part called **ear**. The base feature is a one-sided protrusion off the TOP datum, 10 X 20 with a thickness of 5. We will add an ear to one edge of this base feature.

Once again, the command is not in the ribbon, so go to the ***Command Search*** box and enter "Ear". Then select

> ***Ear > Variable | Done***

Pick on the top surface of the block as the sketching plane. The feature creation arrow should be downwards into the block. The right sketching reference is the right face of the block. Create the sketch shown in Figure 48 below. For an ear, the sketch must be open (turn on the **Feature Requirements** window in Sketcher) and the sides of the ear must be perpendicular to the part edge where they meet. Accept the sketch.

When prompted, enter the following: the depth of the ear is **0.5**, the bend radius is **2**, and the bend angle is **30°**. The resulting ear feature is shown in Figure 49.

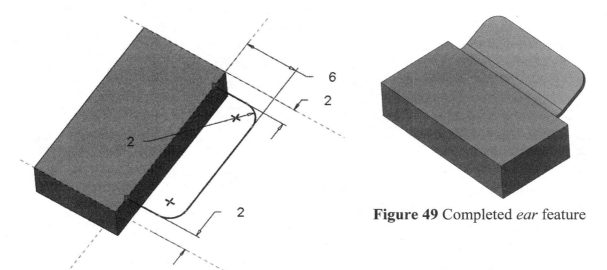

Figure 49 Completed *ear* feature

Figure 48 Base feature and sketch for *ear*

For the ear feature, check out *Edit Dimensions* and *Edit Definition* in the mini toolbar to experiment with other settings (and see some new/old Creo menus!). For example, can the angle be negative? Is there a limit on the bend radius?

This lesson has introduced you to methods to create several special purpose features. Rounds, and especially advanced rounds using multiple round sets and transitions, are a study in themselves. Draft can also be very complicated. The other special features we looked at (ribs, lips, and ears) are not ones you will use a lot, but it is useful to know they are there. For some special cases, these tweaks may come in very handy.

In the next lesson we will look at modeling database functions: pattern and family tables.

Questions for Review

1.   What are the two main operating modes used when creating rounds?
2.   Why are rounds usually created towards the end of the regeneration sequence? When would they be created earlier?
3.   What is the name given to the generic shape of a simple round?
4.   What are the four main types of rounds?
5.   What are the three ways references are used for round placement?
6.   What are the three shapes available for simple rounds?
7.   What determines the shape of the corners where several rounds meet?
8.   How many edges can be contained in an individual round set? Can these edges be intersecting?
9.   How many different ways are there to create a round whose radius changes along its length?
10.   When creating a *Variable Radius* round, does the edge have to be C1 continuous?
11.   How do you create additional points along an edge to specify the round radius in a *Variable Radius* round? Is there a limit on the number of points?
12.   Must a datum curve be used to define a *Through Curve* round or can any edge be used?
13.   What other existing features can be used to determine round size?
14.   What, if any, restrictions are there on the curve used to define the round radius in a *Through Curve* round?
15.   Is it possible to use a *Full* round to remove an L-shaped surface?
16.   Find out how *Edge-Surf* rounds are created.
17.   Does the order of selection of the round sets in a round feature matter?
18.   Can the sets in two separate round features be combined into one feature?
19.   How many round sets can be included in a round?
20.   Find out how Creo determines the default transitions.
21.   Name four types of round transitions.
22.   How do you change a particular transition?
23.   What happens if you use a closed sketch to define a profile rib?

24. What happens if you use an open sketch to define a trajectory rib?
25. What happens if the end vertex of a trajectory rib does not meet a part surface?
26. What happens if you try to attach a 2 unit thick rib to a 1 unit diameter cylinder?
27. Why might feature creation order be an issue if using trajectory ribs that contain both draft and rounds?
28. Can rib features be patterned?
29. Define the following terms:
    neutral curve
    draft surface
    draft reference
    hinge
    draft angle
    pull direction
    split draft
30. How is the pull direction determined if the Draft Hinge is determined by (a) a datum plane, and (b) a tangent chain?
31. What happens if the hinge plane does not intersect the draft surface?
32. What is an undercut? Is there anything in Creo to prevent you from making one?
33. What is a lurking danger if you create drafts early in the regeneration sequence? HINT: Think about choosing reference directions for features.
34. On what type of surfaces can you NOT create draft features?
35. What are the five items needed to define a lip?
36. What restrictions apply to the sketch of an ear?

Project Exercises

We will create four more parts for the cart project. Three of these are pretty routine. The fourth one, the cargo bin, has the highest feature count for an individual part and will utilize many of the functions introduced in this lesson. We'll do the easy parts first. On the hubcap, you can use a function in Sketcher to produce sketched text (see *Sketch > Text*). Protrude this through the hubcap and then cut it off using a revolved cut parallel to the upper curved surface.

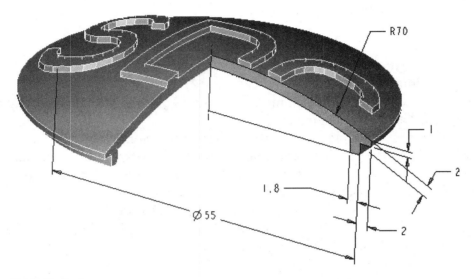

PART: *hubcap*

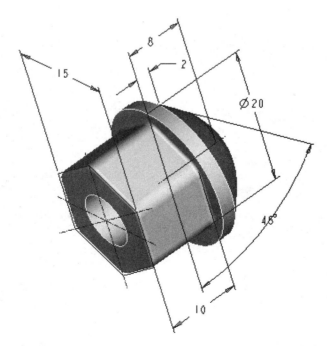

PART: *lugnut*

The figure below shows the cargo bin and some of its interesting features. Plan your feature creation order carefully for this part! It is basically a solid protrusion that has been shelled out. The side walls have a 2° draft. Before you shell out the base feature, add the rounds to the outside edges using advanced rounds and transitions. The shell thickness is 5mm. There are a couple of sweeps, some patterned holes and ribs, and some simple rounds.

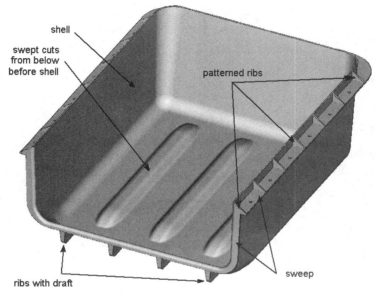

PART: *cargo*

Here are the dimensions for the base feature (a blind, both-sides protrusion) and the corner rounds. This is a case when you would probably put draft on the exterior vertical surfaces of the part prior to forming the rounds and shelling it out.

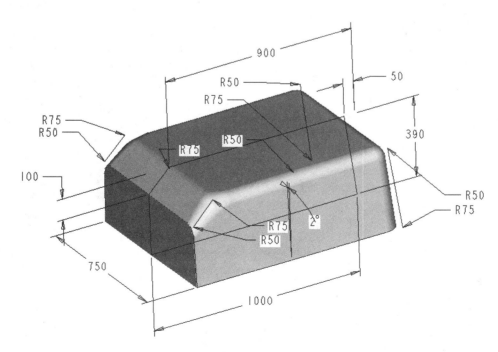

Here are the dimensions for the top sweep and the hole patterns for the cargo bin. For the ribs along the side of the bin, create the pattern leader using an offset **Make Datum** from the end of the bin, say 10mm. The ribs are 5mm thick. Then pattern the ribs (7 ribs, increment 150mm) and change the offset of the pattern leader to half the rib thickness (the rib can't hang out over the end!).

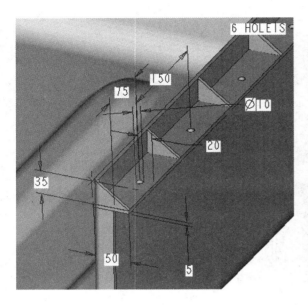
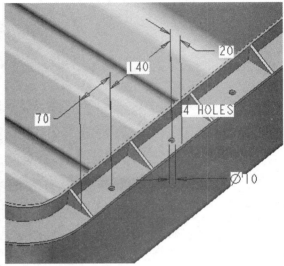

There is also a pattern of ribs on the bottom that look like the figure at the right. These also have draft.

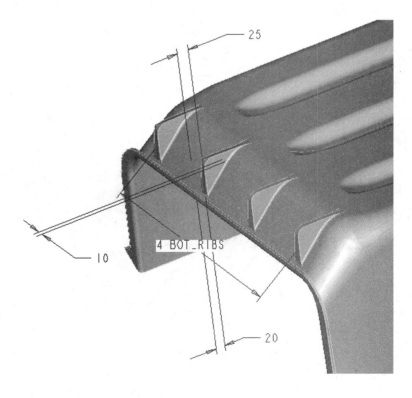

The spring on the side wheels is basically a simple helical sweep. Some added features at each end are required to attach the spring to the bracket and wheel mount.

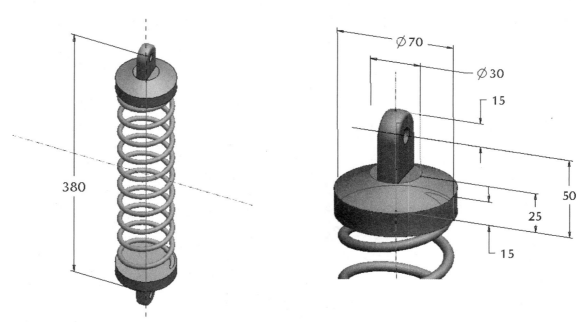

PART: *spring*

IMPORTANT: For an exercise in Lesson 8, you need to create this spring with a single dimension (shown as 380 in the figure on the left) that controls the overall length of the part between mounting holes. One way to do this is to create a datum curve with this length as your first feature. Then, create the end (basically a revolved protrusion with a couple of cuts and a hole) shown in the figure on the right. Mirror this to the other end, then add the helix. The critical point is that the length of the helix should be determined (using references) by the overall dimension of the part. You should include a relation that calculates the pitch of the helix based on the number of turns (a fixed number equal to 10).

This page left blank.

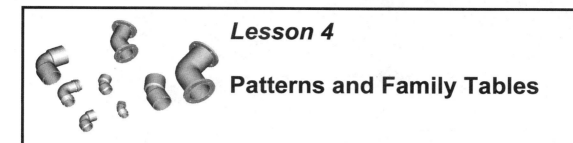

Lesson 4

Patterns and Family Tables

Synopsis

Patterns of features using dimensions and relations; uni- and bi-directional patterns; pattern tables; Reference and Fill Patterns; creating and using family tables for parts

Overview

You will often come across situations where entities or features must be repeated numerous times in a model or assembly. In a part, the easiest way to create multiple copies of a feature is by using the *Pattern* command. This has a number of options and is very flexible. We will look at four types of patterns (**Dimension, Table, Reference, Fill**). Many of the tools for creating patterns of features in parts also apply to creating patterns of components in assemblies. We will deal with component patterns a bit later and stick to features for now. Since pattern creation does not have a complete preview function, you may sometimes be confronted with a failed pattern when it is first accepted. So, we will look at some common causes of pattern failure and how to prevent that.

If you are producing a series of parts that contain variations of a central set of base features, then you can control the geometry of each part *instance* using a simple entry in a table defined in a master part, called the *generic*. This can save you a lot of repetitive part creation, not to mention disk space. Furthermore, members of a part family defined by a table can be freely substituted in an assembly. For this reason, family tables are used extensively in part libraries. In this lesson, we will look at the basics of creating the generic part, the family table, and some issues that involve parent/child relations in the generic part and its instances.

Advanced Patterns

The creation of simple patterns of features was introduced in the first Tutorial. Only one pattern type (**Dimension** patterns) was discussed there. In this lesson we will look a bit deeper into the options available with the other pattern types.

To review, a pattern is a multi-copy operation where a feature (the *pattern leader*) is duplicated in a number of *instances* in a prescribed way. The instances do not have to be identical copies but can change in size and shape as well. Patterns can be created of both placed features (holes and datum points, for example) and sketched features (protrusions, cuts). Patterns can also be made of several features at once, arranged in a *group*.

The four main types of patterns are the following:

1) Dimension pattern - To create this, you choose one or more dimensions defined for the pattern leader that you want to increment or modify between instances or copies. Patterns can be unidirectional or bidirectional as illustrated in Figure 1. The variation between instances can be based on constant increment values or driven by relations. Patterns do not have to be planar. In the exercises below, we will use angular increments (a geometry that seems to cause people a lot of trouble).

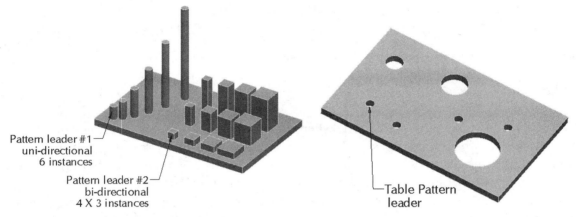

Figure 1 Uni- and bi-directional patterns driven by dimensions of pattern leader

Figure 2 Table pattern of holes

2) Table pattern - This is used to create an irregular pattern of instances using a set of tabulated values for chosen dimensions. This is typically used to create hole patterns, as in Figure 2.

3) Reference pattern - This is created by referencing a feature to the leader in a previous pattern. The new feature can automatically be copied to all members in the initial pattern. If the initial pattern changes (for example, the number of instances changes), the pattern referencing it also changes.

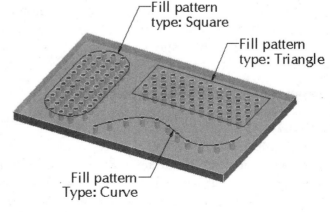

Figure 3 Fill patterns inside bounded areas or along specified curves

4) Fill pattern - This feature can produce the patterns shown in Figure 3, plus others. For these, a simple shape is duplicated to fill a prescribed area using one of several built-in geometric grid patterns or placed along a specified curve or edge.

To pattern multiple features at the same time, they must be placed in a group (they must be contiguous on the model tree for this). If this is done manually, the special *Group Pattern* command must be used. All other details are the same as for creating a pattern of a single feature.

Regeneration Options

A critical aspect of pattern creation is the selection of the regeneration option: *Identical*, *Varying*, or *General*. This option determines the level of freedom you have over the size, references, and intersection of the instances. Identical patterns have the most restrictions and therefore give the least freedom but are the quickest to regenerate. At the other extreme, general patterns give you full freedom but take longer to regenerate. The differences are as follows:

- ▸ *identical pattern* - instances must all be the same size and shape (same edges), use the same reference surfaces, and cannot intersect
- ▸ *variable pattern* - instances can vary in size and shape and use different reference surfaces but cannot intersect each other
- ▸ *general pattern* - instances can vary in size and shape, use different references, and intersect each other

The three types of patterns are created the same way. The regeneration option is determined by a setting in the pattern dashboard. The default is **General**. If the conditions permit, some efficiency is gained by selecting either **Identical** or **Variable**. If a pattern fails, a frequent solution is to upgrade the regeneration option:

Identical ➡ Variable ➡ General

until the pattern properly regenerates. For group patterns, the only option is **General**.

The most important thing to remember about patterns is that, when you are creating the pattern leader, you should be planning ahead to figure out how (or if) you are going to use the pattern leader's dimensioning scheme to construct the pattern. For example, if you want to create an identical unidirectional pattern across the part, there should be only one single horizontal and/or vertical dimension to be incremented to locate the instances. If you have more than one dimension between the feature and the part in either direction but only increment one of them, the instances will not maintain their shape. See Figure 4 and, although both pattern leaders use identical sketches (a 1.0 unit square) for the protrusion, note the differences in the dimensioning schemes and choice of dimensions to the front edge used to create the pattern. In fact, the pattern on the right in Figure 4 cannot be created as an identical pattern (since the instances are changing size); it would have to be created as a varying pattern.

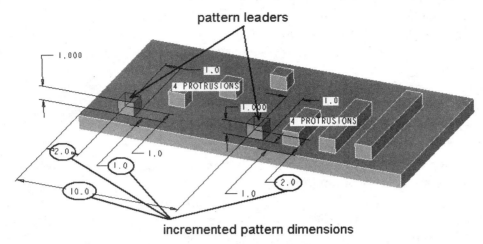

Figure 4 Different dimensioning schemes and choice of dimensions to increment in the pattern is important!

Dimension Patterns

In this section, we will start by reviewing the basic pattern creation procedure for a simple uni-directional pattern. This will give us an opportunity to look at the regeneration options and recovery methods for failed patterns. We will then look at creating radial patterns and the use of pattern relations. We will follow that with the creation of a bi-directional pattern of grouped features, with relations.

Create a new part **pattern1** as a one-sided protrusion off the TOP datum as shown in Figure 5 (sketch constraints not shown). Then create a blind hole (depth 40, diameter 20) located 50 from the left end. This is the feature to be patterned.

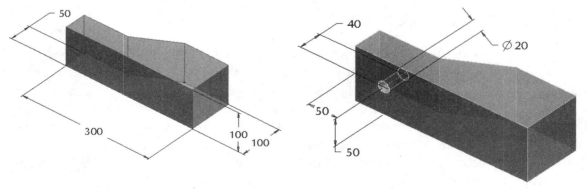

Figure 5 Base feature for part **pattern1**

Figure 6 Blind hole on front surface

We can launch the hole pattern in several ways:

- Use the **Pattern** command in the ribbon (this works object/action or action/object) OR
- Highlight the hole and select **Pattern** in the mini toolbar OR
- Select the hole in the model tree and select **Pattern** in the mini toolbar.

Choose one of these. You will get to the **Pattern** dashboard. On the upper row, the first pull-down list indicates that the default is a **Dimension** pattern. The next four fields are in two pairs, corresponding to pattern directions 1 and 2. For each direction, the first field indicates the number of instances in that direction (the default is 2) while the second indicates the number of dimensions being incremented. You will recall that both location and/or size dimensions can be varied in a pattern. Our first pattern will be in direction 1 only. Click on the horizontal dimension (50) to the center of the hole. A pop-up box appears where we can enter the increment to use for the selected dimension. The default increment is the current dimension value. Press ***Enter*** to accept this (don't middle click, since that accepts the pattern definition as complete). In the dashboard, for direction 1, enter the number of instances as 5.

Before we accept the pattern, have a look in the slide-up panels. In the first panel, **Dimensions**, there are two collectors which will store the selected pattern dimensions and their increment values for the two possible pattern directions. To the right in the dashboard, several tabs are grayed out due to the type of pattern we are creating.

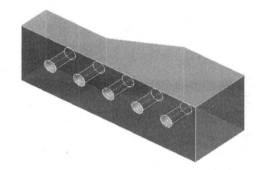

Open the **Options** slide-up panel. This shows the default regeneration option is a **General** pattern. Change this to ***Identical*** (to save regeneration time - this is really important for Fill patterns). The large black dots indicate the

Figure 7 First pattern created (**Identical**)

pattern instances. You can remove an individual instance without disrupting the pattern by clicking on a dot. Click again to get it back. Middle click to accept the pattern feature. The pattern is created as in Figure 7.

Let's do some experimenting with this pattern. First, let's cause the pattern to fail. Open the pattern in the model tree and select the first hole (the pattern leader). Using ***Edit Dimensions***, change the hole depth to **75** and ***Regenerate*** the part[1]. The pattern fails - a Notification window indicates that some features, shown in red in the model tree, failed to regenerate. Have a look around the Notification window, then close it. The problem is that some of the holes go all the way through the block, while some do not, and one only partly makes it through the inclined

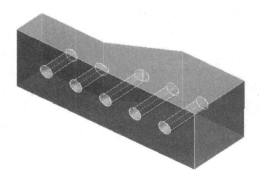

Figure 8 Pattern modified (**Varying**)

surface at the back. Thus, the hole shapes are not identical. We need to promote the pattern to the next level up in the regeneration options list.

---

[1] Remember that regeneration is automatic if ***Auto Regenerate*** is turned on either in the **Operations** group or the RMB pop-up.

Select the **Pattern** in the model tree and in the mini toolbar select *Edit Definition*. This opens the pattern dashboard. In the **Options** slide-up panel, select *Variable* and then accept the feature. The pattern looks like Figure 8. Observe that the first four holes do have quite different shapes.

Let's try something a bit different to see if the pattern leader definition has an effect on the regeneration option. In the model tree, select the Pattern and in the RMB pop-up select *Delete Pattern*. This removes all the pattern copies but keeps the pattern leader. Edit the definition of the hole and change it to a *To Next* hole then create the pattern. Does this also need to be *Variable*? What happens if you define the hole using *Through All* as the depth specification? It is interesting that *Blind*, *To Next*, and *Through All* behave differently (different holes in the pattern fail). Modify the model so that it appears as shown in Figure 9 using the **Through All** and **Variable** options.

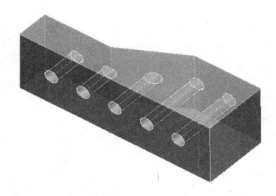

Figure 9 Pattern #3 - holes all the way through the part

As a last experiment, select the Pattern in the model tree and select *Edit Definition*. Check that the direction 1 item collector is highlighted. Hold down the CTRL key and click on the hole diameter dimension to add it to the collector. Set the value **10** for the pattern increment. Confirm this by opening the **Dimensions** panel. The direction 1 collector should list two dimensions (hole location and hole diameter). Check that the **Options** value is set to *Variable*. Accept the feature. Once again, the pattern fails. The reason is not given explicitly, but you can probably guess what the problem is based on what happens with the first four instances. With a diameter increment of 10, the holes are growing large enough that eventually they will overlap and intersect each other. This is not allowed for a Variable pattern. We will have to promote the pattern to General.

Pick the **Pattern** in the model tree, then select *Edit Definition*. Open the **Options** slide-up panel and select *General.* Accept the feature. The failure has been resolved and the pattern looks like Figure 10.

Observe that only the last two holes have intersected. This pattern would actually regenerate as a Variable type with the same parameters if there were only 4 instances. The problem doesn't happen until the fifth instance appears. This gives a clue to an important failure recovery technique with patterns - try using smaller size increments or fewer instances!

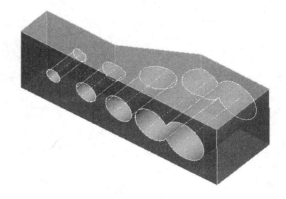

Figure 10 Pattern #4 (**General**)

In the next exercise, we will create a radial pattern by incrementing an angular dimension. We will use this exercise mostly to explore pattern relations.

Using Pattern Relations

We can set up relations to manipulate the dimensions involved in a pattern. In an exercise in the Creo Tutorial, a pattern of holes in a flange (a bolt circle) was created using relations to set the spacing between instances and location of the leader based on the number of holes. Those relations were defined at the part level.

We can also create relations specifically related to the pattern feature and contained within the feature definition. These are stored at a different place from the part relations, and utilize some new built-in symbols (not available at the part level) to give added flexibility.

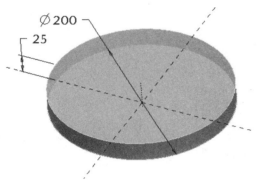

Figure 11 Part to receive radial pattern

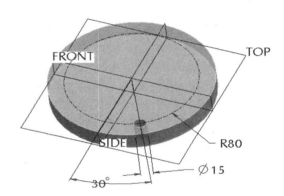

Figure 12 Hole placed using *Radial* option

Start a new part called **pattern2**. Create the base feature shown in Figure 11 as a one-sided protrusion off the TOP datum. Don't forget a geometry point in the sketch to create and axis. Then, create the single ***Through All*** hole using a radial placement as shown in Figure 12. This references the axis of the protrusion and the RIGHT datum plane.

Create a unidirectional pattern of the hole by incrementing the angle dimension. Use an increment of **60°** and set the number of instances to **6**. Change the regeneration **Option** to *Identical*. The pattern will appear as shown in Figure 13.

What we want to do next is to allow the pattern to be automatically modified based on the number of instances. To do this, we need to know the symbolic name of the parameter that stores the number of instances in the pattern. Use

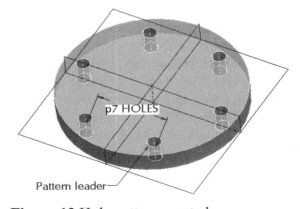

Figure 13 Hole pattern created

the mini toolbar for the Pattern entry in the model tree. Select the ***Edit Dimensions*** command, and the parameter will be displayed. Use ***Switch Dimensions*** (or just mouse-over) to find out its symbolic name (**p7** in Figure 13). Make a note of this name for later.

In the first Tutorial, the symbolic name was used in some relations defined at the part level to control the pattern increment and the pattern leader location. We will do something a bit different here.

Select the pattern in the model tree, then select ***Edit Definition***. Open the **Dimensions** slide-up panel. In the Direction 1 collector, select the angular dimension listed there. Just below the collector, check the box to **Define increment by relation**. Immediately below that, select the ***Edit*** button.

The **Relations** editor window will now open up. A separate information window lists a number of system symbols that can be used in the pattern relations:

> **memb_v** - dimension of resultant position of pattern instance (this will probably depend on either idx1 or idx2)
> **memb_i** - increment for each instance
> **lead_v** - dimension of the leader for the selected dimension
> **idx1** - index for instance in first direction (pattern leader index is 1)
> **idx2** - index for instance in second direction (pattern leader index is 1)

These symbolic names are not available in relations defined at the part level. Some possible uses of these symbols might be:

```
/* a variable spacing for each instance (10, 20, 30, ...)
memb_i = 10 * idx1
```
or
```
/* a sinusoidal variation in instance location
memb_v = lead_v + 10 * sin(idx1 * 180 / p3)
```

The possibilities are endless! Note that memb_v and memb_i cannot be used in the same relation (this wouldn't make sense anyway).

For equally spaced instances in the hole pattern, we want to specify the member increment parameter (**memb_i**). In the Relations window, enter the relation shown in Figure 14, being careful to use your symbolic name on the right hand side for the number of pattern instances. Accept the relation with ***OK***. Accept the feature.

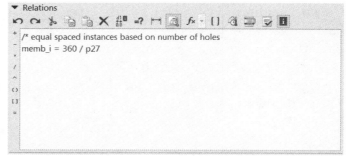

Figure 14 Creating a relation for a pattern dimension

To verify that the pattern relation is working correctly, edit the pattern and change the number of instances to 10. While you are doing that, change the angular dimension of the pattern leader to **0**. ***Regenerate***. This yields the pattern shown in Figure 15. The first hole aligns with the SIDE (or RIGHT) datum.

You can do some very interesting things with pattern relations. Select the pattern again in the model tree and pick ***Edit Definition***. Holding down the CTRL key, add the hole radial dimension (R80) to the direction 1 collector. Then, in the **Dimensions** panel collector, select the new dimension, check the option to use a relation, and pick the ***Edit*** button. Enter the multi-line relation[2] shown in Figure 16. Observe that this relation uses the pattern index, **idx1**. Without looking ahead, can you predict what this will do? Follow the logic for the first three pattern members (the pattern leader has idx1 = 1).

FRONT
TOP
SIDE
Pattern leader

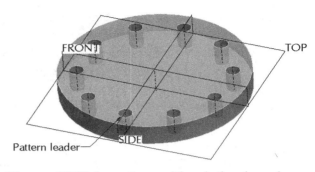

Figure 15 Hole pattern with relation based on number of instances

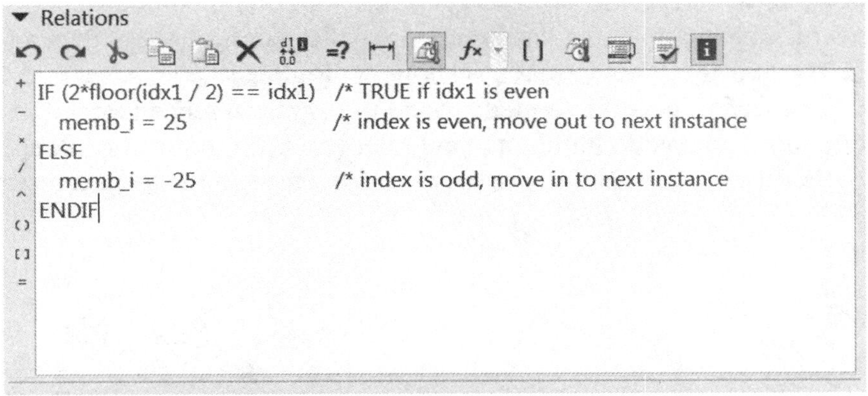

Relations
IF (2*floor(idx1 / 2) == idx1) /* TRUE if idx1 is even
memb_i = 25 /* index is even, move out to next instance
ELSE
memb_i = -25 /* index is odd, move in to next instance
ENDIF

Figure 16 Relation to control the radial placement of the hole

As usual, after you have accepted the new pattern, you should check to make sure the relations work as expected. Try changing the number of instances to any other even number (like **12**, see Figure 17). What happens if you use an odd number?

Pattern leader

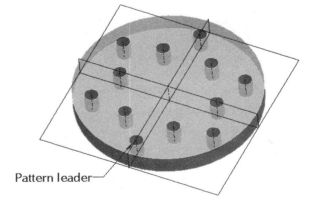

Figure 17 Pattern relation controlling radial placement of hole

---

[2] The *floor(...)* function returns the largest integer less than the argument enclosed in brackets. For positive arguments, this basically truncates anything after the decimal point.

Editing Pattern Relations

You should remember that the relations above are defined for (and stored with) the specified pattern dimension, that is, not at part or feature level. It is easy to forget this later and lose track of where the relation went. A quick way of accessing pattern relations is to use (in the **Model Intent** group overflow)

> *Relations*

Select *Pattern* in the **Look In** pull-down list box (note other options: part and feature) and click on one of the holes. Or, selecting (in the other list box) one of the dimensions governed by a relation will bring up the relation itself in the new dialog window. You can now edit the relations as usual. When you are finished, save the part and remove it from the session.

Bi-directional and Group Patterns

In this exercise, we will combine several aspects of pattern creation: grouped features, radial patterns of sketched features, bi-directional pattern definition, and using pattern relations. The part we are going to make is shown in Figure 18. This is a thin-walled cylinder with a pattern of cuts distributed around the circumference (we will call this the cylindrical direction) and axially along the cylinder (axial direction). Although not really obvious from the figure, the width of the cuts also increases in the axial direction.

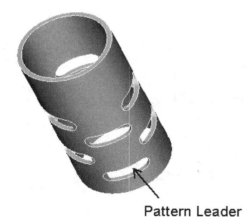

Pattern Leader

Figure 18 Completed part with patterned cuts

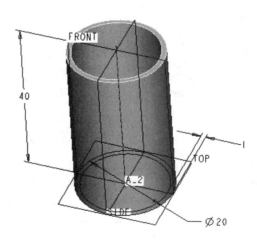

Figure 19 Base feature for *pattern3*

Create a new part called **pattern3**. Start by creating a hollow cylinder as a blind protrusion off the TOP datum plane. The outer diameter is **20** and the wall thickness is **1**. The height is **40**. See Figure 19.

Now we create the pattern leader. This cut will be sketched on a datum plane created tangent to the cylinder. In order to pattern this feature in the cylindrical direction, we want to have an angular dimension around the cylinder axis to be associated with the cut. We could do this by creating a number of datum planes, one of which is located by an angle dimension, grouping the associated datums and the cut, and then finally creating a group pattern. This would produce a large number of datum planes in the model. Instead, we will employ a couple of make datums in the pattern leader. The make datums will be visible in the model tree, but hidden in the graphics window.

Select the *Extrude* command. In the RMB pop-up menu, select *Define Internal Sketch*. We do not have a suitable sketching plane so we will create one on-the-fly. To see where we are going, see Figure 20.

Select the *Datum Plane* command in the drop-down at the right end of the dashboard. For the datum plane references, select the axis of the cylinder and (with CTRL) the RIGHT datum. Enter an angle 30° and accept the datum. This will be the angle we can use to create the pattern. The datum plane is labeled DTM1 in Figure 20.

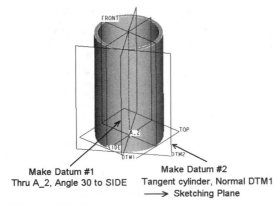

Make Datum #1
Thru A_2, Angle 30 to SIDE

Make Datum #2
Tangent cylinder, Normal DTM1
⟶ Sketching Plane

Figure 20 Constructing make datums for the cut

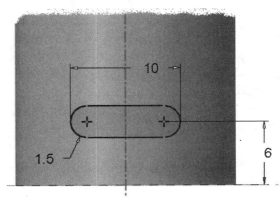

Figure 21 Sketch for the cut

We still do not have a suitable sketching plane, so once again select the *Datum Plane* tool. For the new datum, select DTM1 and the outer surface of the cylinder. In the references collector in the DATUM PLANE window, change the references to be *Normal* to DTM1 and *Tangent* to the cylinder surface. The new datum is labeled DTM2 in Figure 20. Select this as the sketching plane and TOP as the Top sketching reference, then activate Sketcher.

Create the sketch shown in Figure 21. There is a vertical centerline aligned with the vertical datum DTM1, so that a symmetry constraint will be created. Note the dimensioning scheme. The vertical dimension will be used to create the pattern in the axial direction. Complete the sketch and set the feature depth to *To Selected*, then pick on the inside surface of the cylinder. Don't forget to change the *Add/Remove* toggle. Accept the feature. See Figure 22. Notice there is no sign of the make datums to clutter up our view! In the model tree if you open the extrude feature, you will see both DTM1 and DTM2 listed but set to **Hidden**. It is curious that they are listed *after* the Sketch.

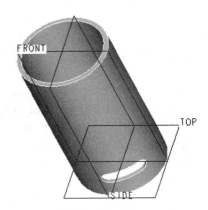

Figure 22 Finished cut - the pattern leader

Bi-directional Patterns

We want to create a bi-directional pattern with our Direction 1 around the circumference of the cylinder and Direction 2 along the axis. Select the extruded cut in the model tree and in the RMB pop-up menu, select *Pattern*. All the dimensions involved in the feature appear, including the 30° angle dimension that was used in the construction of the make datums.

For Direction 1, click on the angle dimension to DTM1 and enter an increment of 120°. In the dashboard, set the number of instances for the first direction to **3**. Now click in the dimension collector for Direction 2. Click on the height dimension of the cut (currently 6) and enter an increment of **6**. Holding down the CTRL key, click on the radius dimension on the sketch (R1.5) and enter an increment of **0.25** - this will make the cuts get wider. Finally, still in the collector for the second direction, CTRL-click on the angle. We want to stagger each row, so enter an increment of 30°. The number of instances in the second direction is **4**.

Figure 23 Completed pattern

To confirm the pattern set-up, open the **Dimensions** slide-up panel. There should be one dimension (the angle) listed in the Direction 1 collector (increment 120), and three dimensions listed in the Direction 2 collector (increments 6, 30, 0.25).

If all is well, accept the pattern. The part should look like Figure 23. Open the model tree to see how this is structured there, including the instance numbering scheme. Only the pattern leader contains the make datums. Save the part.

Some More Pattern Relations

For this part, we want to set up pattern relations so that the cylindrical spacing will be adjusted if we change the number of instances in that direction. We also want each row to line up with the gaps between the cuts in the row below. This will require two simple relations. Each will involve the symbolic name for the number of instances in the cylindrical direction.

In order to find the symbolic name for the number of instances in each direction of our pattern, we first have to create the pattern with a fixed increment and given number of instances. This was done above. To find the symbolic name, select the pattern in the model tree, and then *Edit Dimensions*. The leader dimensions will appear along with the symbols giving the number of cuts in each direction. Select *Switch Dimensions* to show symbolic dimensions. In Figure 24, the symbol p12 refers to the cylindrical direction, and symbol p16 refers to the axial direction. **Your symbolic names will most likely be different from these, so make a note of them.** (Two dimensions have drag handles. What do these do? Come back later to check these out!)

Now we can set up our relations. Select the pattern in the model tree, then *Edit Definition*. Open the **Dimensions** slide-up panel. In the Direction 1 collector, select the dimension (current increment 120). Then check the box "Define increment by relation," then select the *Edit* button. Enter the following relation (using your symbolic name on the right):

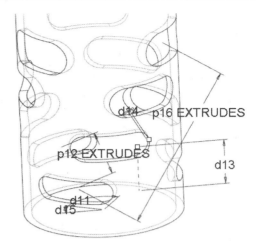

Figure 24 Finding symbolic names

$$\texttt{memb\_i = 360 / p12}$$

Leave the editor with *OK*.

Back in the **Dimensions** panel, in the collector for the second direction, select the same angular dimension (currently 30). We want the angular offset for the next row to be half the spacing of the first row. Check the box below the collector, then *Edit*. Enter the relation for this direction and increment (remember to use your own symbolic name for the number of instances, p12)

$$\texttt{memb\_i = 180 / p12}$$

Close the editor and accept the redefined feature.

Try out the new relations using *Edit* and changing the number of cuts in each direction. One variation is shown in Figure 25. What happens if you try to create too many cuts in the cylindrical direction? in the axial direction? What happens if you increase the length of the cuts from 10 to 15 with 4 cuts around the circumference? Can you predict and/or explain what troubles might arise before you try these changes? There should not be many problems with these variations in the pattern because of the default regeneration option (which is ...?).

Figure 25 Pattern driven by relations

We are finished with this part, so save it and then remove it from the session. So far, we have created patterns of regularly spaced features. In the next section, we will explore a powerful way of creating a pattern of arbitrarily spaced or located features.

Table Patterns

Patterns controlled by dimension values and/or relations generally result in regularly spaced instances. Relations allow you to locate instances using formulas or expressions. For example, with appropriate relations you could pattern holes around an elliptical opening (although there is an easier way of doing this with a Fill pattern). For a pattern of features that have truly irregular locations, we need a new tool - a pattern table.

Start by creating a new part **pattern4**. Create a base plate (20 X 15 X 2 thick) and single circular protrusion (diameter 2, located as shown in Figure 26). Since we are going to pattern the protrusion, make sure it has dimensions as shown giving its location from the left and front faces of the base plate.

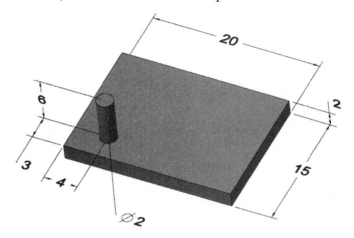

Figure 26 Base feature and pattern leader

Modifying Dimension Symbols

For what we are about to do, it is very handy to have symbolic names for dimensions. Highlight the cylindrical protrusion and select **Edit Dimensions**. Click on the horizontal dimension to the center of the boss. In the Dimension dashboard in the **Name** field, enter a new name "**boss_x**". Similarly, select the dimension that gives the vertical location, and call it "**boss_y**". You may have to select the **Switch Symbols** to see these. See Figure 27. These names for the dimensions will appear in the pattern table, and are much more meaningful there than d15 and d16, or whatever.

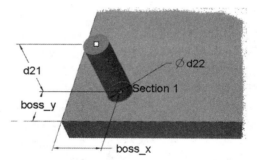

Figure 27 Symbolic names

Creating the Pattern Table

Now we're ready to set up a pattern table. Select the cylinder, then pick the **Pattern** command. In the dashboard pull-down list, change the pattern type from **Dimension** to **Table**. Read the message window. Click on the horizontal dimension (boss_x or 4) and (using CTRL) the vertical dimension (boss_y or 3). In the **Tables** slide-up panel, which is now active, change the name of the active table to BOSSES. Note in passing that this option allows us to define several different data tables for the feature, only one of which will be active at any time. Close the Tables tab and back in the dashboard, select the **Edit** button. We are now in the Pro/TABLE editor. In this table we can enter values for the chosen dimensions for each instance in the pattern. Each row in the table corresponds to one instance and contains the following columns:

> **idx** - the instance number, starting from 1. This does not include the pattern leader, which is automatically created with the table.
> **boss_x** - the horizontal location
> **boss_y** - the vertical position

The pattern leader is shown in row 9 in Figure 28, with its corresponding dimensions in parentheses. For any row below this, if you want to use the pattern leader's dimension for that column, just enter an asterisk, "*". The formatting and alignment of this table is not great - pretty much all you can do is change column widths. Enter the data shown in Figure 28, and then select *File > Exit*. In the pattern dashboard, set *Options > Identical* and accept the feature.

R7	!	Table name BOSSES.		
R8	!			
R9	!idx	boss_x(4.0)	boss_y(3.0)	
R10	1		16 *	
R11	2		10	7.5
R12	3 *			12
R13	4		16	12
R14				

Figure 28 Pattern table for *bosses*

We now have the 5 bosses shown in Figure 29 according to the (x,y) positions specified in the table. Note that although the four corner bosses could have been created with an incremental dimension pattern, we would not be able to create the 5th boss (idx = 2 in the table) in the center that way (without some tricky relations).

A common use of pattern tables is to create irregular patterns, for example mounting holes in a plate.

Figure 29 Pattern created using table

How can we change the locations of the bosses? Double click on one of the bosses in the pattern, say the back right boss (number 4 in the pattern table). Change its height (6) to **10**, and its horizontal position (16) to **12**. *Regenerate* the part. Notice that the height dimension is the same for all bosses, since it was not included in the pattern table. The horizontal location has changed because it is unique to that instance.

For another way to change the locations, select *Pattern Table* (in the **Pattern** pull-down list in the **Editing** group).

A new TABLES window opens up. Highlight the table entry "BOSSES" in the tree. In the icons at the bottom of the window, find the one (third from left) that will "Edit the Selected Table." Observe that the boss_x value for instance idx=4 has changed. Change the horizontal position of the last boss back to 16 in the table. Leave the window with *File > Exit*. The cylinder will move back when you regenerate. You can go back into the pattern table and add new instances to the pattern just by adding new rows to the table. Pretty easy! We now have two ways to modify entries in the pattern table: directly on the model, or by entering the table itself. Find out what is the difference between **OK** and **Apply** in the Tables window.

Before we continue, change the height of the cylinders back 6. **Regenerate** the part.

Reference Patterns

A reference pattern involves a second pattern laid on top of a first pattern. The second pattern will "copy" any references used in creating the first pattern.

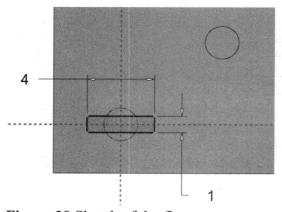

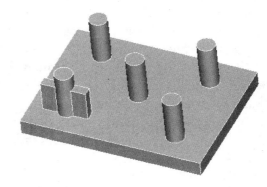

Figure 30 Sketch of the fin

Figure 31 Fin added to pattern leader

Create a small solid protrusion at the location of the cylinder pattern leader, as shown in Figure 31. We'll call this a fin. It is crucial that this new feature reference only the geometry of the cylinder pattern leader. To do that, using Intent Manager you can select just the cylinder as a reference. Then create vertical and horizontal centerlines. This will allow a symmetry constraint to be set up for the rectangular sketch, shown in Figure 30. Create the protrusion with a blind height of **4**.

Now we want to put a copy of each fin on the other members of the boss pattern. Select the fin and then **Pattern**. In the dashboard, the type **Reference** may have been selected automatically. If not, do that now and accept the feature. That's all there is to it! The fin duplicates itself onto all the instances in the reference pattern. See Figure 32.

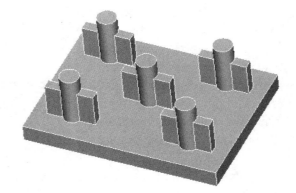

Open up the model tree to see how these two patterns appear there. How is this different from if we had created a group

Figure 32 Fins created using reference pattern

(boss+fin) and patterned the group? What would be the functional difference of this arrangement?

Let's edit the cylinder pattern table. In the **Editing** group, select:

> ***Pattern Table***

Double-click on "BOSSES" to open the table editor. Select the last row in the table and remove it using ***Edit > Delete***. Exit from the editor and TABLES dialog window and ***Regenerate*** the part. This removes the cylinder and the fin. Similarly, adding new entries to the cylinder pattern would also automatically produce new fins.

The important thing to remember here is that the leader in the second pattern should reference only the pattern leader in the reference pattern.

To see the effect of the "*" entries in the pattern table, double click on the cylinder pattern leader and change the horizontal dimension (boss_x) to **6**, and the vertical dimension (boss_y) to **8**. ***Regenerate*** the part. The two pattern instances that use "*" also change with the leader. Also, notice that since the fin pattern is a general type, the fin features can intersect (even though the cylinder pattern is set to **Identical**).

Save the part and erase it from the session.

Fill Patterns

The **Fill** pattern is used to create a regular (equally-spaced) grid-like arrangement of a simple feature within some bounded area. All instances must have the same size but they can intersect. The boundary can be defined by object edges or datum curves. Several grid geometries are available, including spiral and polar grids. You can control grid orientation, member spacing, clearance from the boundary, and presence/absence of individual members of the grid. A special "grid" lets you pattern members along a curve.

Let's experiment with some simple fill pattern options. Start a new part called **pattern5** and create a protrusion (20 X 15 X 2 thick). Or, ***Save a Copy*** of part **pattern4**, open it, and delete the cylinder pattern and leader.

We can create the fill boundary either before or after we launch the Pattern tool. We will do it before - create the sketched curve on the top of the base feature something like the one shown in Figure 33. (Hint: use the ***Offset*** tool in Sketcher, with the ***Loop*** option). Now create a single through-all hole (diameter 1) somewhere inside the boundary. The location of this pattern leader will be the origin of the pattern grid.

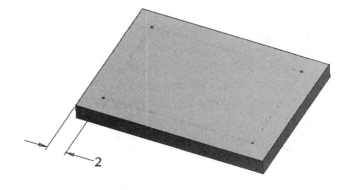

Figure 33 Sketched datum curve to serve as Fill pattern boundary

Select the hole, and pick ***Pattern***. In

the pattern type pull-up list, change the type from Dimension to *Fill*. This makes a number of changes in the dashboard (starting from the left):

- a collector for selecting existing boundaries. The boundary collector is active, so just click on the sketched curve. This activates the rest of the dashboard.
- the grid type (default **Square** - check out the other types available)
- instance spacing within grid
- minimum distance to boundary
- grid angle
- radial grid spacing (for **Circle** pattern only)

A preview of the default fill appears on the model using large black dots. Notice the location of the pattern leader. Observe the white drag handles. What do they do? Change the type to *Hexagon*, and the spacing to *2*. The pattern should look something like Figure 34.

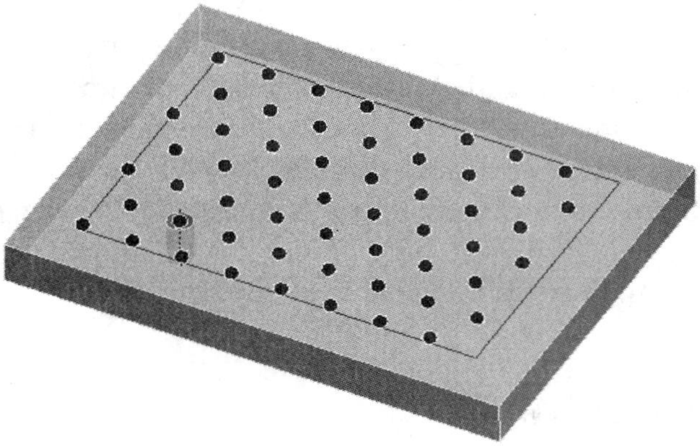

In the dashboard **Options** panel, notice that we can select the usual regeneration options. Select *Identical* here to reduce regeneration time. What does *Use Alternate Origin* do?

Figure 34 Fill pattern with *Triangle* grid

(HINT: select the option, then pick a corner vertex of the bounding sketch.)

You can selectively remove individual instances from the pattern by clicking on the preview dots. A removed instance changes to white. Clicking on it again brings it back. Notice that you can remove the pattern leader. Starting from the triangle pattern of Figure 34, removing selected instances can produce the hexagonal pattern shown in Figure 35.

Experiment with these fill pattern options. If you pick the pattern type *Curve*, you can create a pattern of holes along the defining boundary

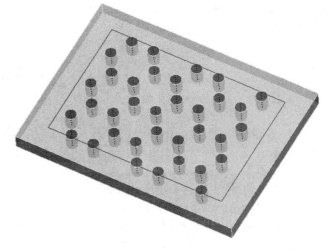

Figure 35 Fill pattern with removed instances

(Figure 36). Unlike a fill pattern, the location (offset) of the pattern leader from the curve is critical. The curve has a default **Start Point** and the offset of the pattern leader from this point is maintained around the pattern. To change the **Start Point**, you may have to unlink the curve sketch (in the dashboard select ***References > Unlink***) then edit the curve to place the start point at a desired vertex (see RMB pop-up in Sketcher once the vertex is chosen). You may also have to move the hole pattern leader so that it is on the start point of the curve. Unlike a fill pattern, you cannot remove the pattern leader from the pattern. You can specify the number of instances, which will be equally spaced along the curve.

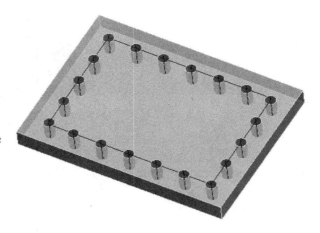

Figure 36 *Curve* pattern with 20 equally-spaced instances

Try setting a very small member separation. If it is too small, the pattern will fail. Why? How can you fix it?

We have finished looking at patterns. Since a lot of design involves repetitive features, a good knowledge of patterns can be very useful. You should experiment with these options as much as you can. We will now move on to another very useful function that can decrease the amount of repetitive labor in creating similar models.

Family Tables

As mentioned in the overview to this lesson, family tables are to parts what pattern tables are to features, with the major difference being that family table members are stand-alone parts. Typical uses for family tables are to construct series of similar-shaped parts like bolts, fittings, housings, gears, and so on. To create a family of parts you do the following:

- create a *generic* part (in a pattern this would be called the pattern leader),
- create a table to specify which dimensions, features, and/or parameters will vary among the *instances* or individual copies of the part,
- edit the table to create new instances by specifying an instance name and associated values for the selected dimensions, features, and/or parameters

In this lesson, we will create a family table that describes some of the parts shown in Figure 37. These elbow fittings have variable diameter, variable angle, variable wall thickness, and can have one of several different end conditions (flange, hose, pipe). All of these variations are contained in a single generic part.

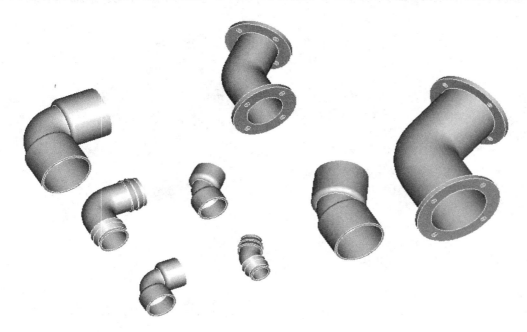

Figure 37 Instances of a part created using a family table

Creating the Generic Part

We will construct the generic part using a sweep to create the base feature. Start a new part called **elbowg**. First, construct a sketched curve to serve as the trajectory for the sweep. Sketch this in the TOP datum plane. The sketch is shown in Figure 38.

Use the datum curve as a trajectory to construct a simple solid sweep whose section is a circle of diameter **50**. Then use the *Shell* command to remove the end faces, leaving a thickness of **2.5**. At this point, your part should look like Figure 39.

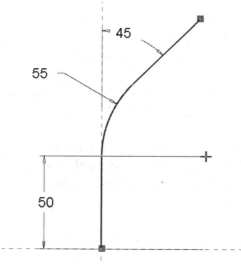

Figure 38 Datum curve sketch

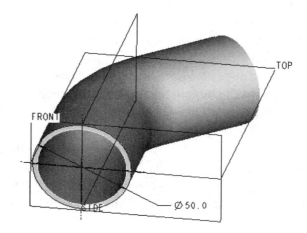

Figure 39 Elbow base feature (sweep along datum curve) plus shell

In the family table, we will enter the elbow angle, pipe diameter, and shell thickness as dimensions to control the instances. Also, so that these dimensions will control other dimensions in the part, we will set up some relations for the length of the straight section and the radius of the elbow. For both these operations, it will be handy to change the dimension symbols. We did the same thing when dealing with pattern tables earlier in this lesson. Double-click on the associated features (the datum curve, sweep, and shell), highlight each dimension, and change the dimension name. For the angle dimension, enter a new name "**angle**". For the diameter dimension enter a new name "**diam**". See Figure 40. Change the symbolic name of the shell thickness dimension to "**thick**".

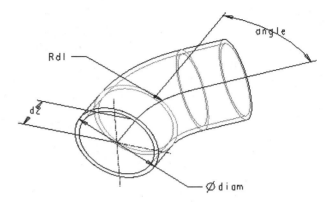

Figure 40 Named dimensions for use in family table

Now we want to enter some relations. A quick way to do this is to double-click on the datum curve. Double-click on the curve radius dimension (55) and enter "**1.1*diam / 2**". Creo asks if you want this added as a relation to the part; select *Yes*. This makes the radius of the elbow slightly larger than the radius of the pipe. Why do we want to do this? The answer is so that the sweep will always regenerate for any value of the elbow diameter. If we don't modify the trajectory radius this way and enter a very large pipe diameter, the sweep would fail.

Double click on the length dimension for the datum curve (50) and enter "**diam / 2**" and add this as a relation. Now the straight ends of the elbow will change their length depending on the pipe diameter.

Try out the new relations. First, regenerate the part to see the shape governed by the relations. Then change the angle dimension to **30** and the section diameter dimension to **20**. *Regenerate*. Try an angle of **90** and a diameter of **60**. When you are satisfied that the relations are working correctly, proceed. Otherwise, use *Tools > Relations* to make necessary corrections.

For the following, change the dimensions back to angle **30** and diameter **20**. Save the part.

Creating the Family Table

In the **Model Intent** group, select

> *Family Table*

This brings up the Family Table editor window, Figure 41. To use it, basically follow the instructions in the window. The family table operates the same as a pattern table: each column will contain entries for a selected parameter, and each row will yield an instance of the part. Unlike a pattern table where instances are numbered, the second column in a family table contains an instance name. The icons across the top are your controls for manipulating the table.

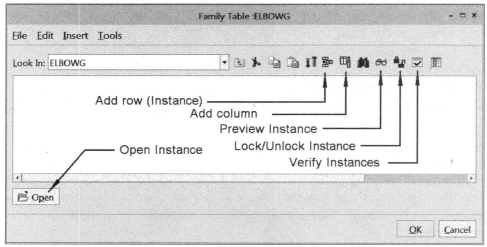

Figure 41 The Family Table editor window (interior text removed)

Click the *Add Column* icon. This opens another window (**Family Items**) shown in Figure 42. Some choices are grayed out at this time. With the *Dimension* radio button selected in the *Add Item* area at the bottom, pick the datum curve, and select the angle dimension. Pick the sweep section and select the diameter dimension. Finally, pick the shell and select the thickness dimension. As each dimension is picked, it will be added into the item list (note the modified symbolic names we created earlier). Select *OK* when all three are listed.

Back in the Family Table editor, the name of the generic part is indicated along with the generic part dimensions. Each row below the generic is used to create an instance of the part. In the second column of the table, under *Name*, we will enter unique part names for all the instances we want to define. In the 4th and subsequent columns are the identified dimensions we will vary among instances. To use any of the generic part's dimensions in an instance, use the "*" symbol.

In the pull-down menu, select

Insert > Instance Row

or use the *Add Row* icon, or select any cell in the first row and hit the *Enter* key. In the new

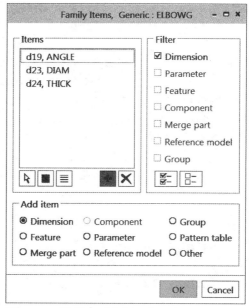

Figure 42 Specifying items to include in the family table

row, enter **E-60-40-3** as a new instance name in the second column. In the angle column, enter **60**. In the diameter and thickness columns enter **40** and **3**, respectively.

Enter another part instance **E-30-20-2**. In the dimension columns enter angle **30** (or use "*"), diameter **20** (or "*"), and thickness **2**. Using the "*" symbol is handy to get the value from the generic. Beware of using the "*" - this will have ramifications discussed a bit later. The completed family table is shown in Figure 43.

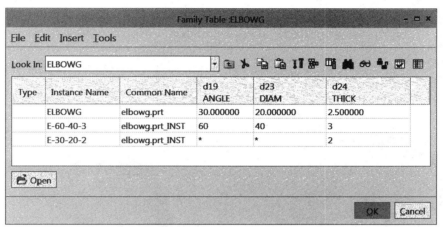

Figure 43 Completed family table

Verifying the Family Table

Near the right end of the icons, select the **Verify Instances** icon. This will validate all the instances in the current family table. A small window will open and allow you to choose which instance(s) to check. Highlight both the instances and select **Verify**. If all has gone according to plan and all instances will regenerate successfully, you should see the entries "Success" beside each one. A failure would obviously require us to go back to the generic part to debug the model. If you have a very large model and/or family table, the verify could take some time. **Close** the verify window.

Accept the family table with **OK** and save the part in the working directory.

A summary of the verification results is written to a file, in this case **elbowg.tst,** in the current working directory. Open this file with your system editor. It contains the name of each instance and the regeneration result (Success or Failure).

Examining Instances

On the screen, we still have the generic part (this is indicated at the bottom of the graphics window). To view the geometry of an instance, open the family table again in the **Model Intent** pull-down menu and highlight the instance name in the second column of the family table and select the toolbar icon

Preview

This opens another small window which displays the instance (this is a dynamic view window). Close this window.

Another way to look at an instance (or if you want to do additional work on it, like add more features) is to highlight the instance name and select ***Open*** (bottom left of family table window). For example, do this with the instance *E-60-40-3*. A new part window opens. Note that in the lower left corner of the screen you are notified that the part being displayed is an instance, and the name is given. This window is now active, meaning that you could further modify the instance by adding features and so on. We will discuss modifying the instances a bit later. Come back later and find out what happens if you try to save an instance of the part[3]. For now, close its window with ***Close*** (in the Quick Access toolbar). Bring up the other instance of the part. The two instances are shown in Figures 44 and 45.

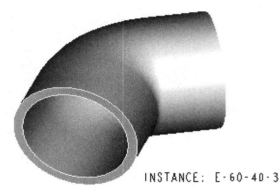
INSTANCE: E-60-40-3

INSTANCE: E-30-20-2

Figure 44 An instance in the family table

Figure 45 ... Another instance

Adding Features to the Family Table

So far, all we have included in the family table are the variable dimensions for the instances. We can also specify optional features in the generic part which can be included in the instances. These features must be regenerated in the generic, and we have the option of keeping them or suppressing them in each of the instances. We will use this capability to create alternative end conditions for the elbow.

This can get complicated due to parent/child relations. In the family table, we identify specific features for presence/absence in the instance. If these features have children in the generic, then we are also implicitly controlling these children in the instances, because if the parent is suppressed, all its children will be suppressed as well.

Furthermore, all the optional features must be able to independently coexist in the generic part. If the regeneration or suppression of one feature in the family table interferes with the references of another, then these references will have to be redefined.

---

[3] **HINT**: Set your *config.pro* file with the following option:
   display_full_object_path   **yes**
and observe the title area at the top of the Creo window when you have an opened instance. Also, have a look at the idx file in your working directory.

We will create two alternative ends for the elbow - a flange and some ribs for a hose fitting.

First, we'll create the ribs for the hose fitting (Figure 47). The ribs are created using a 360° revolved protrusion, sketched on the RIGHT datum. The sketch is shown in Figure 46 and the completed ribs are shown in Figure 47. HINT: when setting up references for this sketch, use *X-sec* in the **Sketcher References** dialog window to create a reference on the outer surface of the pipe where it intersects the sketching plane. Otherwise, you will have to dimension the diameter of the sketch and use a relation to connect it to the "diam" parameter value.

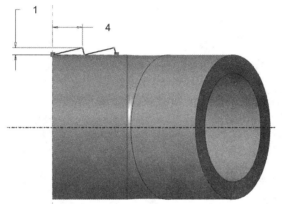

Figure 46 Sketch for the hose fitting (Sketcher constraints not shown)

Figure 47 Hose fitting on generic part

Now we'll create the flange as a 360° revolved protrusion. But first, we want to make sure that the flange does not reference the ribs in any way by suppressing the ribs. For the flange, use the RIGHT datum as sketching plane, and create the sketch shown in Figure 48. IMPORTANT: Use the FRONT datum as a sketching reference, not the vertical surface of the end of the pipe; also, use X-sec as before to create a reference on the outer surface of the pipe. See Figure 49 for the completed flange.

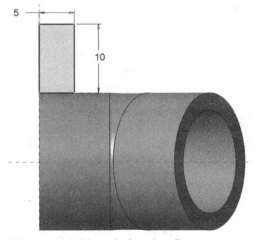

Figure 48 Sketch for the flange

Figure 49 Flange on generic part

Resume the ribs into the generic part, so that both the flange and the ribs are available[4]. You might reposition your screen so that both features are visible. Note that the flange completely covers the rib closest to the end of the elbow. Rename the flange and rib features to "**flange**" and "**hose**". This will make entries in the family table a little easier to follow (just like renaming dimensions).

Now, in the **Model Intent** group, select

Family Table > Add Column

and check the option *Feature* in the **Add Item** area. Select the flange and hose features in the model tree. Middle click three times.

We have added a couple of columns[5]. In the generic part, both features are present ("**Y**"). Edit the two instances as shown in Figure 50. Note that we are also changing the name of each instance ("F" = flange, "H" = hose). In the last two columns (these may be in the opposite order from what is shown in Figure 50), the Y and N entries control the presence of that feature in the instance. For the flanged elbow, we want to suppress the hose ribs, and vice versa. Use *Verify* to check the definitions of the two instances, then *Preview* them. Leave the table editor with *OK*.

Type	Instance Name	Common Name	d19 ANGLE	d23 DIAM	d24 THICK	F503 FLANGE	F429 HOSE
	ELBOWG	elbowg.prt	30.0	20.0	2.5	Y	Y
	E-60-40-3-F	elbowg.prt_INST	60.0	40.0	3.0	Y	N
	E-30-20-2-H	elbowg.prt_INST	*	*	2.0	N	Y

Figure 50 Features added to the Family Table

Child Features in the Generic Part

We want to duplicate the flange/hose fittings at the other end of the elbow. We could make these independently, but this would require additional columns in the family table. A more elegant way is to create these as children of the flange/rib features created above. There are a number of ways of doing this. Several of these require the presence of a datum plane created *Through* the datum curve and *Normal* to the TOP datum at the back of the elbow. You might like to enter *Insert Mode* and create this feature immediately after the sketched curve. This datum plane might also come in handy when we get to assembly mode, since we will have similar datum planes at both ends of the elbow.

To create the duplicated end fitting features, some possibilities are the following:

---

[4] This may be tricky. If the geometry is not what you expected, check the sketching references in the flange when the ribs have been resumed. You may have to modify some of the sketch references and constraints.

[5] You can change column widths by dragging on the column separators.

▸   create a *Copy > Move > Rotate*, and set up a relation for the rotation angle
▸   create a *Copy > Paste Special > Dependent copy | Advanced ref configuration*
▸   create a new feature, being careful in Sketcher to reference the existing previous
     fitting to establish the parent/child relation (or use relations to connect the
     dimensions of the copied feature to the original)

Choose whichever method makes the most sense to you (or try all three!). As you are
creating each feature, you may want to suppress some features to avoid inadvertently
setting up a parent/child relation you don't want. Name these two features **hose_copy**
and **flange_copy**.

The completed generic part with both
features on both ends of the elbow is shown
in Figure 51. Now would be a good time to
check on the parent/child relations in the
part. You can do this by individually
suppressing either the hose feature or the
flange feature. The copies (and nothing
else!) should be suppressed automatically
with them.

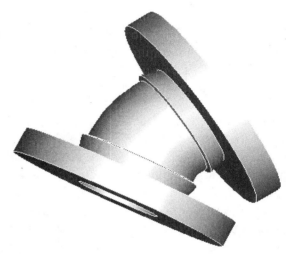

With our new features added to the generic,
go to *Family Table > Verify* to validate the
table. With the *Preview* command, call up
each of the instances we have created. If
everything is correct, you should get the
parts shown in Figures 52 and 53.

Figure 51 Generic part with both rib and
flange features regenerated

INSTANCE: E-60-40-3-F

Figure 52 An instance in the family table

INSTANCE: E-30-20-2-H

Figure 53 Another instance in the family
table

We have now essentially completed the creation of the generic part. Save it.

Bring up a directory listing of the working directory in the Browser window. The generic
part **elbowg.prt** is listed. Depending on how your Browser is set up (see the Browser
Tools pull-down menu, and check the option *Show Instances*), the instances may also be
listed. If you go to Windows Explorer and look at the directory, you will observe that the
instances are not saved as separate files. You will also notice another file with the same
root name as your working directory and the extension *idx*. This is the instance index file

for this directory. Open this up in your system editor. It contains a listing of all instances defined for any generic part in the current directory that has a family table defined. The idx file is updated whenever you save the generic part. There are additional functions that reference the idx file. See the command *File > Instance Accelerator* in the pull-down menus. For further information, see the on-line documentation.

With the generic part saved, close all part windows, then *Erase Not Displayed*. We want to see what will happen when we open the generic into an empty session.

Manipulating Parts Containing Family Tables

Select *File > Open* or choose the icon on the toolbar. Although the instances do not exist in the working directory, you will see an entry in the **File Open** dialog window for each instance. As part of the instance name, you will see the generic part, as in <*elbowg*>. This information was obtained from the index file. *Cancel* this window without selecting a file.

Open the Navigator and select your working directory. In the Browser window, select one of the instances, say *E-30-20-2-H*, and open it. When the part is brought in, you will see the INSTANCE label on the screen. Notice the part description at the top of the window - it shows that the instance is the active part, but the generic is what has been loaded. Check out the *File > Manage Session > Object List* command - it also indicates that both the generic and instance parts are in session. Check out the model tree contents for the instance - no sign of the flange features. Try the *Family Table* command. This does not show us the same data as before. In fact, we could now create another family table with the hose elbow treated as the generic. Nested tables! This is something to experiment with later on your own. For now, close the family table window.

Change the thickness of the shell in this instance to **1**. As you do this, you are notified that the shell thickness is table driven, and you must confirm modification to the family table entry back in the generic part. *Regenerate* the instance and save it. (What actually gets saved here?)

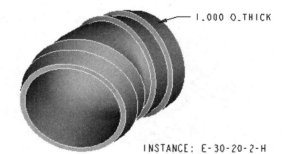

Close all the part windows, and remove everything from the session with *Erase Not Displayed* in the Home ribbon.

Figure 54 Modifying a table-driven dimension in an instance

Now, open the generic part. If you select *File > Open* and select the generic part *elbowg.prt*, you will see the window shown in Figure 55. Examine the contents under the two tabs. Using the **By Column** options can help you locate a family member with a specific set of parameter values. Highlight "The generic" in the *By Name* tab and select *Open*. Then select *Family Table*. The table is shown in Figure 56. Note that the thickness value that we changed manually for the second instance has been modified in the table, but the generic has not been affected (thickness still 2.5).

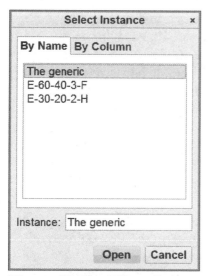

Figure 55 The *File > Open* dialog for a part containing a family table

Type	Instance Name	Common Name	d19 ANGLE	d23 DIAM	d24 THICK	F503 FLANGE	F429 HOSE
	ELBOWG	elbowg.prt	30.0	20.0	2.5	Y	Y
	E-60-40-3-F	elbowg.prt_INST	60.0	40.0	3.0	Y	N
	E-30-20-2-H	elbowg.prt_INST	*	*	1.0	N	Y

Figure 56 Modified table in generic part due to modified value in instance (Figure 54)

What happens if you modify a dimension in an instance that is not driven by the family table? Open up the other instance, *E-60-40-3-F*. (HINT: select the instance in the family table, then select *Open* at the bottom of the window.) Change the flange thickness to **2**. *Regenerate* the instance. See Figure 57. Go to the window containing the generic part (use *Window* in the Quick Access toolbar to select the generic part). The flange dimension in the generic part has also changed as shown in Figure 58. Note that the rib closest to the end of the elbow is now slightly visible in the generic part.

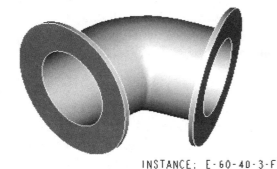

Figure 57 Modified flange dimension in instance (not table driven)

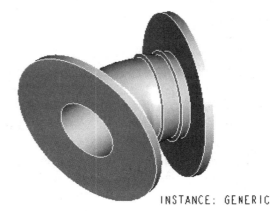

Figure 58 Modified dimension in generic (not table driven)

Thus, we have the following things to note:

- Modifying a table-driven dimension in the instance changes the table entry in the generic part for *that instance only*
- Modifying a non-table driven dimension in the instance changes the generic part (and therefore *all other instances using that dimension or feature*)

If you want to modify a non-table driven dimension in an instance without affecting the generic part, you must create an independent part file. Do this using *Save A Copy* when the instance is active. Select and make active the window containing the instance *E-30-20-2-H*. Save this, providing a unique name for the new file, say **E3020TEST**. See the message window ("Copied part will not be table driven.") - you might have to scroll back a few lines. Open the working directory again and you should see this as a new (independent) part file. Back in Creo, open this new file. It will appear without the INSTANCE label on the screen. Open the model tree - there are no suppressed features in this part so the flange feature (and copy) have been left behind. Change the shell thickness to **3**. Open the windows for the previous instance *E-30-20-2-H* and the generic *elbowg*. Neither of these has been affected by the change in the independent part file E3020TEST.

Locked Instances

When a family table is created, to prevent inadvertent changes to the table driven dimensions (like the shell thickness above), you can put a "lock" in the first column of the family table. Do that now in the family table by selecting the instance *E-30-20-2-H* and using the **Lock/Unlock** icon. Switch to the instance, and try to modify the thickness of the shell. Creo will inform you in the message area that the dimension is locked. If a number of people are using the same generic parts (like bolts, for example), this prevents them from making changes to the table definition that could affect the other users.

Adding Features to an Instance

Additional features can be freely added to an instance. Given that instances are not stored in separate part files, you can probably guess that parent/child relations might get pretty complicated in a table-driven family. To reduce the chance of problems, you should try to put optional features in the generic part near the end of the regeneration sequence, and not add too many more features in each instance.

Let's add a pattern of holes to the front flange on instance *E-60-40-3-F*. Bring in the instance, and create the first hole using a diameter dimension scheme. Make a straight hole with a diameter of **5**, and select a **Thru Next** depth specification. Place the hole on the front flange at an angle of 45° from the TOP datum. In the data entry box for the placement diameter dimension, enter "**diam+10**". This will make sure the holes are outside the diameter of the pipe. Create a pattern of 4 holes using an increment of **90°**. See Figure 59.

Switch windows to the generic part. The hole pattern is not displayed on the model. In the model tree, you should see the hole pattern is suppressed in the generic part. Open the family table to see the extra column that has been added. This gives a clue to (perhaps) an easier way to create features that will only appear in some instances - create them first in the instance itself.

INSTANCE: E-60-40-3-F

Figure 59 Bolt hole pattern created in instance.

Switch back to the instance and select *Save*. Remember that this does not store the instance as a separate part.

Erase all objects in your session.

This concludes our discussion of family tables for now. We will return to the use of family table-driven parts in a later lesson, to see how these can be used efficiently in an assembly.

This lesson has introduced you to several different methods to create patterns of features. This included an introduction to relation, fill, and table-driven patterns. We also looked at using family tables to define variations of similar parts with different dimensions and features. As usual, we have introduced the main ideas for these topics and there is considerably more to learn by browsing through the on-line documentation and experimenting on your own. For example, it is possible to include a pattern table within a family table so that different instances can have different patterns of the same feature. This might be useful to specify a bolt hole pattern that will vary from one part instance to another. You can also construct families of families; the Creo BASIC part library is structured this way. As usual with Creo, careful planning up front is required to efficiently use these advanced functions, and no planning up front is an invitation to disaster.

In the next lesson we will return to examining some of the functions in Creo to create single features and feature groups using User Defined Features (UDFs).

Questions for Review

1. The first feature in a pattern on which all the others are based is called the _____.
2. What are the four main types of patterns? What are the main differences between these?
3. What is a common cause of the failure of an **Identical** pattern? How would you attempt to fix it?
4. What happens to the pattern leader when you select **_Delete Pattern_**?
5. What is your only regeneration option if you are going to create a pattern of features that will intersect each other?
6. In a bi-directional pattern, does the second direction always result in a pattern at 90° to the first direction?
7. What are the built-in symbols that we can use in creating pattern relations? How are these defined? Give an example of the use of each symbol.
8. Do the following relations result in the same pattern?
    ```
    memb_v = 10*(idx1 - 1)
    memb_i = lead_v + idx1*10
    ```
9. In a dimension-driven pattern, what happens if the increment is negative?
10. From the part level, what is the shortest command sequence you can use to find out the relations that have been used in creating a pattern?
11. What command will toggle the display of dimensions/symbols?
12. Why can there be multiple definitions of *memb_i* in a single feature?
13. How can you rename the dimension symbols in a part?
14. What is the difference between symbols idx1, idx2, and idx?
15. What are two ways to modify the values in a pattern table?
16. What happens if you modify a feature dimension that is not included in the pattern table?
17. What is the primary requirement for using a reference pattern?
18. What are the three main steps required to create a family table?
19. What command do you use to validate the entries in the family table?
20. What is the main complication when including features in the family table?
21. Why should you rename features to be used in a family table? What happens if you do not?
22. Can you rename a feature after it has been included in the family table? What entry occurs in the family table in that case?
23. What is contained in the instance index file? How is it identified? Where is it stored? Can you have more than one?
24. How can you change a dimension in an instance without affecting the generic part?
25. How can you prevent accidental changes to table-driven dimensions in the generic part?
26. What will happen if you modify a table driven dimension in the instance, that is indicated by "*" in the family table in the generic part?
27. Can you create an instance in a family table that contains a feature that is suppressed in the generic part?

Project Exercises

The project parts in this lesson involve patterns, pattern tables, and family tables.

Start by adding the horizontal holes to the lower side frame. Create the pattern leader 320mm from the end and then a pattern table, where the pattern contains the dimension of the hole from the end of the frame. The pattern table is shown at the right.

!	Table name HOLE_LOCNS.	
!		
! idx	d27(320.0)	
1		380.0
2		470.0
3		530.0

✱ *FRAM+LOW-RGT*

The next part to make is the main side wheel. The base feature is a revolved protrusion with the key dimensions shown in the figure. Add additional features such as a hub and a pattern of cuts to make the spokes. The 4 holes in the center are a simple radial pattern. Estimate the location now, and you can modify the pattern diameter later to match the axle we will build in the next lesson. Note that the bolt holes have a 45° chamfer. This will mate with the conical surface of the lug nuts made in Lesson 3. Try using a *Ref Pattern* for the chamfer.

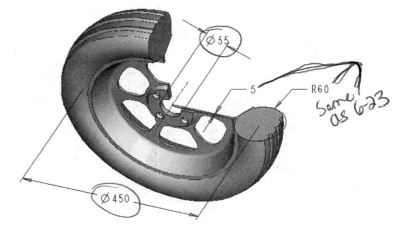

PART: *wheel_1*

We'll now make two parts that involve family tables. The first is the square tubing to be used in the frame. This is shown in the figure on the right. The family table is shown on the next page.

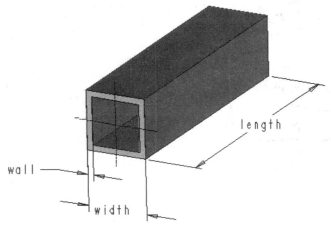

PART: *tubing*

! Generic part name: TUBING			
! Name	d2	d1	d0
!	length	width	wall
!			
! ================ ================ ================ ================			
! GENERIC	100.0	25.0	2.5
T25X125	125.0	25.0	2.5
T25X325	325.0	25.0	2.5
T25X375	375.0	25.0	2.5
T25X75	75.0	25.0	2.5
T25X825	~~TT50~~ 825	25.0	2.5

The last part in this lesson is a hex-head shoulder bolt. We won't bother with the threads on this. However, notice the lip feature on the top surface (see the previous lesson). The length of the bolt is set in the family table.

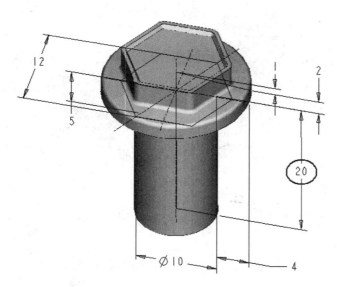

! Generic part name: HEX_BOLT	
! Name	d3
!	length
!	
! ================ ================	
! GENERIC	20.0
H20	20.0
H30	30.0
H40	40.0
H50	50.0

PART: *hex_bolt*

Blind 0.50

Add taper

30°

Lesson 5

User Defined Features (UDFs)

Synopsis

Creating and using User Defined Features; *Standalone* and *Subordinate* modes; *Independent* and *UDF Driven*; using family tables; dimension display modes; patterns of UDFs.

Overview

Since the beginnings of CAD a few decades ago, a fundamental principle has been to never reinvent the wheel! Any time a reasonably complicated or common model (or parts of one) is produced, it makes sense to reuse it in other jobs rather than to recreate it each time. Furthermore, in a corporate environment, there may be particular standards and practices in modeling which are desired. This is the essence of User Defined Features (UDFs). In the former case, UDFs are feature definitions that are stored in separate files and can be brought into any part (and most times assemblies as well). The UDF can be used over and over, thus reducing the need for repetitive modeling activity. In the second case, a library of UDFs can be maintained such that anyone working at a particular site can access and use the same geometric elements. This promotes good practice and standardization and more productive modeling.

This lesson will introduce the main aspects of creating and using UDFs. We will go through three exercises, starting from something very simple, to illustrate the main options and principles. The first exercise contains a UDF composed of a single feature. The second exercise is considerably more complex and involves several features and a family table that is contained within the UDF. The third exercise shows how a single UDF definition file can be used to control the geometry in many other parts simultaneously, even after the UDF has been placed in these other parts. Finally, we will have a look at the UDFs included in the Creo UDF Library.

Introduction to User Defined Features

One of the more confusing aspects of UDFs, at least initially, is the meaning of terms. So, a good place to start is some definitions. These are consistent with the Creo on-line documentation, so after you finish this lesson you can comprehend the documentation a bit better. The important terms are given below. Other terms relate to options that will be presented as we create and use the UDFs in the exercises later on. We will discuss them as they arise.

Definitions

These definitions are illustrated schematically in Figures 1 and 2.

Original model - This is the model where the UDF is created. One or more features in the model are identified as belonging to the UDF. The original model can be the source for several UDFs, depending on how the features are chosen. The model can be an actual part (or assembly) under construction, or, more usually, it is a simple base feature to which only the features for a single UDF are added. Depending on how the UDF is created, the original model may have to be present at all times when the UDF is used (see Exercise #3 below). At other times, a copy of the original model will be created (see *reference model* below) from the original. A third possibility is that once the UDF is created, with certain options, the original model can be deleted.

New model - This is the model where the UDF is used. The UDF definition is read from a file and the features are reconstructed in the new model by specifying references, assigning values to variable dimensions, and so on. Note that in the new model, the UDF appears in the Model Tree as a group, even if composed of a single feature, with the same name as the UDF.

***Standalone* UDF** - If a UDF is created so that its connection to the original model is broken, then it is classed as *Standalone*. See Figure 1. The original model is not required in order to use the UDF, and changes in the original model will not affect it.

Creating the UDF   Using the UDF

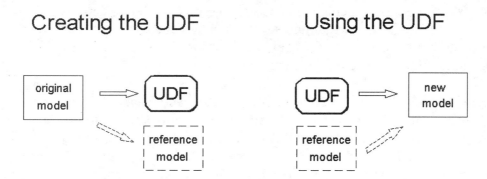

Figure 1 Creating and using a ***Standalone*** UDF

Reference model - When a standalone UDF is created, you have the option of creating a copy of the original model along with the UDF. This copy is called the reference model (Figure 1). The purpose of having a reference model is so that, when the UDF is being placed in the new model, you can view the reference model in a second graphics window. This makes it easier to determine or visualize the references, dimensions, orientation, shape, and so on, of the UDF before it is placed. In order to keep the size of the reference model small and its appearance simple, the original model will usually contain only a base feature and the necessary references to define the UDF.

Subordinate **UDF** - This form requires the presence of the original model whenever the UDF is used. See Figure 2. When the UDF is brought into the new model, the original model is consulted for dimension values and other information required to create the UDF. This essentially takes the place of the reference model defined above and can result in reduced disk space requirements. It also makes it very easy to update the UDF by simply modifying the original model. Whenever a new model is retrieved that uses the UDF, the current definition of the UDF in the original model is used. This means, of course, that wherever the new model goes (for example, to a new site), the original model must go with it.

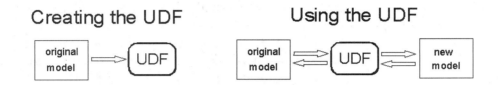

Figure 2 Creating and using a *Subordinate* UDF

There are some other combinations of options during creation and use of UDFs. However, the two modeling scenarios shown in Figures 1 and 2 illustrate the common operating modes.

Invariable **dimensions** - When the UDF is brought into a new model, these are dimensions as they appear in the original model. Each invariable dimension can be set to one of three states when the UDF is applied: **Unlock** (the dimension can be modified in the new model), **Lock** (the dimension can be displayed but not modified), or **Hide** (the dimension will not display and is therefore locked). An example of an invariable dimension that would be locked is the diameter of a standard hole - this would not likely change in the new model.

Variable **dimensions** - When the UDF is brought into a new model, the user is automatically prompted for values for these dimensions. The default values are those contained in the original model. The dimension can be freely modified in the new model. An example of a variable dimension would be the depth of a blind hole.

Now, on to some exercises...

Creating and Using UDFs

As mentioned above, we are going to do three exercises using UDFs, of gradually increasing complexity. These will illustrate both *Standalone* and *Subordinate* operation, as well as some variations and options in the UDF definitions.

Exercise #1: Standalone, Independent

The first example will introduce you to the major steps involved in creating and using a UDF. This will be a simple example so that you can see an overview of the process. The part we want to make is an oval shaped cover plate with six "pins" that mate in sockets in another part. The final cover plate and a closeup view of one pin are shown in Figures 3 and 4. The reason we want to use a UDF for this task is so that the pin geometry can be reused in additional cover plates without going to additional effort in recreating the pin geometry each time.

The pin will be created as a *Shaft* feature[1] so that the only placement reference required is a single datum point in the center of its base. If we used a revolved protrusion, we would have to specify sketching and reference planes. It is a good idea to have the simplest reference scheme possible for the UDF, as there will then be fewer steps involved, less confusion, and fewer chances for errors later on when the UDF is used.

Figure 3 Cover plate with six UDF pins

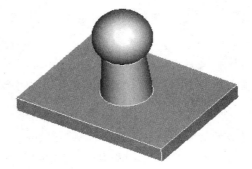

Figure 4 Pin created as *SHAFT* feature (reference part)

Creating the Original Model

Start a new part called **udf_pin_org**. Using *File > Prepare > Model Properties*, set the part units to millimeters (unless that is in your default template). Create the protrusion shown in Figure 5, a 20 X 15 X 2 block. Place a datum point on the top surface near the center of the base feature (do not use a Sketched Datum Point). The dimensioning scheme for the datum point will not affect the UDF. The pin is created as a *Shaft* feature. Use the *Command Search* function for this, since it is not on the ribbon menus. Also, this will involve some of the old-style cascading menus. Use the *On Point* placement option to place the lower center of the stem on the datum point. The sketch for the shaft is shown in Figure 6 (remember that the *Shaft* feature requires a closed sketch and an axis of revolution). The completed original model with the pin is shown in Figure 4.

---

[1] You will need the *config.pro* option
allow_anatomic_features   yes
to be able to create this feature. The default is **no**.

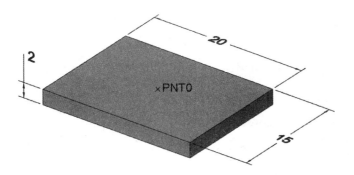

Figure 5 Base feature of original model

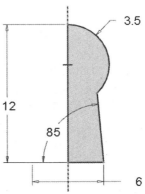

Figure 6 Sketch of *SHAFT* feature

Creating the UDF

To create a UDF of the pin, in the **Tools** ribbon, **Utilities** group, select

> *UDF Library > Create*

You are prompted for the name of the UDF; enter "**cover_pin**". Select the option

> *Stand Alone > Done*

You are now asked if you want to include the reference part. Select *Yes*. We now enter some old style (pre-Creo) menus.

The UDF element definition window appears as shown in Figure 7. At the top of the window are shown the name and type of the UDF. The first thing to do is to select features to be contained in the UDF. The UDF FEATS menu should be automatically brought up with the following defaults highlighted

> *Add > Select*

Pick on the shaft. It highlights in green. Middle click and select *Done/Return*.

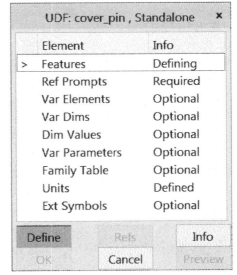

Figure 7 UDF creation elements

The next item in the elements window involves the reference prompts ("Ref Prompts"). What are these? In this case, the only placement reference for the UDF is the datum point (highlighted in blue). When we use the UDF in a new part, Creo will prompt for this reference location on the new part. We can specify the text to be used in the prompt. Keep the text simple but clear. When using the UDF in the new part with the reference model displayed, Creo will also show the reference graphically in the reference model. However, since the display of the reference model is optional when the UDF is used, you will want to have a clear prompt here.

Enter the text "*pin placement datum point*". You can come back later to modify the text if desired. In the SET PROMPT menu, select ***Done/Return***.

This completes the definition of the required elements for a UDF. We have not specified any of the optional elements. In particular, we have not specified any dimensions as being *variable*. All dimensions in this UDF are therefore *invariable*. This means that when the UDF is brought in to a new part, it will contain the original UDF dimensions and not prompt for alternative values in the new part. All users of the pin UDF will then have the same initial values to work with.

In the UDF element window (Figure 7), select ***OK***.

Note the contents of the message window. We are informed that the group *cover_pin* has been stored. The file name for the UDF is *cover_pin.gph*. By the way, the ***Modify*** command in the current (UDF) menu allows us to come back to change any aspect of the UDF definition. At this time, the reference part *cover_pin_gp.prt* is also stored. In the UDF menu, select ***Done/Return***. The storage location for the UDF and the reference part is the current working directory. Once you have the UDF finished and completely debugged (that is, after you have tested it out!), you may want to transfer both these files to a UDF library directory. You can set up the option **pro_group_dir** in *config.pro* to tell Creo to automatically search this library for UDFs. If the directory is accessible to other users, they can use the UDF, too.

We are now finished with the original model, *udf_pin_org*. ***Save*** the part and then ***Erase*** it from the session. You can also erase all the non-displayed objects.

Using the UDF

Start a new part **cover_plate**. Set the units to millimeters. Create a base feature as a simple protrusion using the dimensions shown in Figure 8. Create three datum points on the upper surface of the cover plate according to the dimensions shown in Figure 9. You might like to use a pattern table for this (remember that dimensions in a pattern table can be negative).

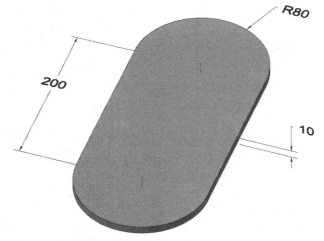

Figure 8 Cover plate base feature

Now we bring in the UDF and place it on one of the datum points. Go to the **Model** ribbon, **Get Data** group, and select:

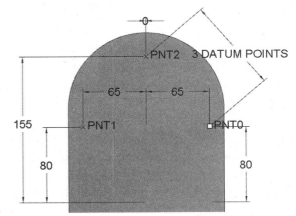

User Defined Feature

The **Open** file dialog window opens up. Depending on how your system is set up, the dialog will either be pointing to your working directory or to a special directory on your system for storing UDFs called the *Features Library*. This may vary according to settings in your *config.pro*[2]. We will

Figure 9 Cover plate datum points

have a look at some of the built-in UDFs contained in this library a bit later in this lesson. For now, select the working directory in this window. The UDF *cover_pin.gph* should be listed. Highlight this item and select *Open*.

A new dialog window opens containing options involved in the placement of the UDF:

Make features dependent on dimensions of UDF (unchecked by default)
> Leave this unchecked to make the feature independent of the original UDF model. The UDF definition *cover_pin.gph* can be changed (or even deleted from the disk), but these changes will not be reflected in the new part. Leaving the UDF independent increases the portability of the new model (since it no longer requires the presence of the definition file) but increases the file size of the new model. Checking this option, in other words making the UDF dependent (also called *UDF Driven*), requires the presence of the UDF definition file whenever the new model is loaded or updated but allows the "keeper" of the UDF library to enforce standard usage among all users. That is, if the UDF definition is modified, these changes will (more or less) automatically occur in any models using the UDF when they are loaded.

Advanced reference configuration (checked by default)
> Setting for how references of the placed UDF will be determined. Leave this checked.

View source model (unchecked by default)
> For the first few times that you use a new UDF, it's a good idea to check this option. Do that now. Even though we provided prompts for the UDF references, it is a good idea to have a display of the UDF source model to be certain what the references are. This is particularly important for a UDF containing multiple features and several references.

Select OK. This opens the reference model showing the UDF. The datum point, which is the only UDF reference, is highlighted in bright blue. A second window, the **User Defined Feature Placement** window, opens which gives us options for how we want to treat the UDF.

---

[2] The *config.pro* option looks something like (your path may be different)
`PRO_GROUP_DIR c:/ptc/objlib/featurelib`

In the **UDF Placement** window, the first tab deals with **Placement** references. The pane on the left contains all references required for the UDF. In this case, only one reference is required; its prompt (that we entered earlier when defining the UDF) is shown on the right side of the window. Pick on the left datum point on the cover plate model (PNT0 in Figure 9) to receive the UDF. The second tab, **Options**, gives us some choices about the size of the feature and how we want to treat its dimensions.

In the **Options** tab, we can either keep the same dimensions or scale the model. We want the UDF in the new model to have the same numeric dimension values as in the reference model. Note that if the units (inches and/or mm) of the UDF and new model are different, the default is to convert the UDF to the new model unit system by scaling it. Come back later and try this out!

Next, we can set an option to determine how the dimensions in the UDF will be accessible in the new part. When we created the *cover_pin* UDF, by default we did not identify any dimensions as being *variable* (we'll do that in the next exercise). Therefore they initially assume the values in the original model. As we will see later, dimensions that are set to *variable* can be changed when the UDF is being brought in. When placing the UDF, we have three options for dealing with existing dimensions as follows:

▸   *Unlock* - converts the invariable dimensions to variable, so that we can later *Edit* them in the new part if desired (this is only available for Independent UDFs)
▸   *Lock* - shows the original UDF dimensions in the new model, but they cannot be modified
▸   *Hide* - uses the original dimensions but will not display them in the new model and they cannot be modified.

We are going to try out each of these on our cover plate. For the first UDF in the new part, select

Unlock (the default)

and middle click. The pin is now placed. Check out how it is indicated in the model tree.

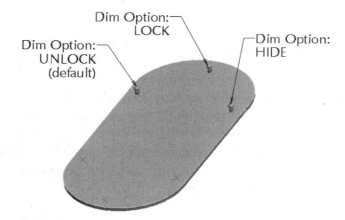

Repeat the above placement procedure two more times, placing the same UDF on two other datum points. Keep all the UDFs **Independent**. For each of these, use a different setting (*Lock* and *Hide*) in the **Option** tab, as indicated in

Figure 10 Options when placing a UDF

Figure 10. When you go to select the UDF in the **Open** dialog window, check out the "In Session" icon (blue screen icon). The UDF is already in session.

Zoom in on each pin, select the pin and using *Edit Dimensions* try to change their dimensions. On the first pin (**Unlock**), change the height to 15 and the diameter of the base to 4 and *Regenerate* the part. For the second pin (**Lock**), you will get a message that the values cannot be modified. If you try the third pin (**Hide**), the *Edit Dimensions* command doesn't even appear.

Delete the pins. Assuming that you used a pattern table for the datum points, use the UDF to make a pin on the datum point pattern leader. In the Options tab, set the scale factor to **2.0**. Edit the point pattern table to create the three other points at the other end of the plate. Then it's an easy matter to make a *Ref Pattern* of this first pin (this takes just three mouse clicks: select pin, *Pattern*, *Accept*) since the defaults are exactly what we want. It should now look like Figure 3 at the beginning of this exercise.

This completes our first UDF example. *Save* the part and remove it from the session. Erase the non-displayed objects as well.

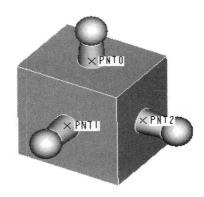

Before we leave this example, note that the UDFs do not have to be created on the same surface or in the same orientation as the source part. Our UDF is more flexible than this. The part shown in Figure 11 was created using the same *cover_pin* UDF. This is because the *Shaft* feature automatically aligns itself to the surface on which it is placed. This is an example of keeping your UDF references simple!

Figure 11 UDF placed on different surfaces

Now, on to something a bit more challenging.

Exercise #2: Multi-feature UDF with Family Table

The basic procedure for creating and using UDFs is very similar to a Copy/Paste kind of operation. However, UDFs can get much more sophisticated (i.e. complicated). In this example we will use some of the options that we omitted in the previous example when creating the UDF: variable dimensions and a family table.

Our goal is to create a UDF that will allow us to easily produce any of the recessed holes shown in the part below (Figure 12). Note that the shape of the recess can be either circular or hexagonal. The diameter of each hole and the size (width or diameter) and shape of the recess will be determined using a family table defined in the UDF. These will be *invariable*. We will create the UDF so that the user can specify the hole placement using a radial dimensioning scheme (distance from an axis and angular placement from a reference plane), and the depth of the recess. The placement dimensions and recess depth will be *variable*. Once created, the UDF group can be easily patterned as shown in Figure 13, using the angular placement dimension.

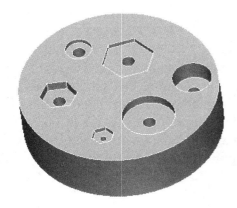

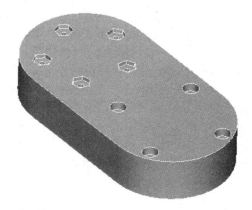

Figure 12 Recessed holes created with UDF *combo_hole*

Figure 13 Patterns of UDF *combo_hole*

Creating the Original Model

Start a new part called **udf_recesshole_org**. Set the units to millimeters. Create a solid protrusion as shown in Figure 14. This is a 200 mm square block, 50 mm thick, centered on the vertical datums. Create a datum axis at the intersection of the FRONT and SIDE (or RIGHT) datums.

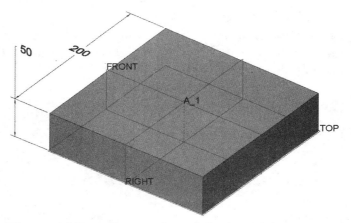

Figure 14 Base feature for original model

Create a straight *Through All* hole with a diameter of **10** using a radial placement on the top of the block. See Figure 15. The reference axis for the hole is the datum axis, and the angular (polar) dimensioning reference is the FRONT datum plane. The same types of references will be required to place the UDF. Place the hole at an angle of **30°** from the FRONT datum. The placement type is *Diameter* (see the Placement panel in the Hole dashboard) and the value is **100**.

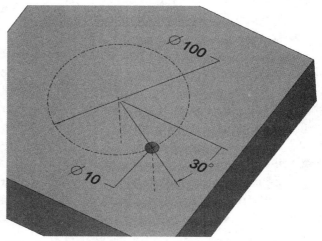

Figure 15 Creating the hole feature

Now create the circular recess as a *Coaxial* hole. Select the axis of the hole and the top of the block for the placement references, and enter a *Blind* depth of **10** and a diameter of **20**. See Figure 16.

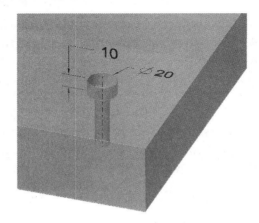

The hexagonal recess will be a bit more tricky because we want to avoid setting up unnecessary references to the part. So far, the only references we have used are the hole axis, the center axis (for the dimension type), the FRONT datum for the angular reference, and the top surface. If we can, we'd like to avoid bringing in additional references (like part edges) since that will complicate the UDF (and possibly not exist in all models).

Figure 16 Creating the circular recess

Also, we want the hexagonal recess to exist independently from the circular recess. The easiest way to ensure this is to *Suppress* the circular recess. Do that now.

Create the hexagon as a *Blind* extrusion feature (don't forget to select *Remove Material*). Use the top of the block for the sketching plane.

IMPORTANT:

For the *Top* sketching reference, create a *Make Datum* through the axis of the hole and the central axis. This ensures that all references required for the hexagon are contained within the UDF. These two axes are already involved in the UDF so we are not introducing unnecessary new references. In addition, we want the orientation of the hexagon to rotate according to the angular placement of the UDF. This will occur because the Make Datum will rotate with the angle specified for the hole. Otherwise, we could have selected the FRONT datum as the sketching reference since it is already a reference in the UDF. Do not add an unnecessary reference if you can avoid it.

The hexagon sketch is shown in Figure 17. Using Intent Manager, the only sketching references we need are the axis of the hole and the *Make Datum*. The easiest way to create this sketch is to draw a circle, right click and select *Construction*, then draw the lines to create the hexagon. Dimension the sketch as shown in Figure 17. Or, even easier, use the **Polygons** tab in the Sketcher *Palette*. You should be able to fully dimension the sketch with the single dimension shown and some combination of constraints.

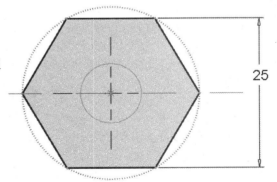

Figure 17 Sketch for hexagonal recess

With a successful sketch, set a **Blind** depth of **5**. Remember that the circular recess has a depth of 10, so we will be able to see both features when they are resumed. Accept the feature.

Now resume the circular recess and you should see all three features (hole, circular recess, hexagonal recess) as shown in Figure 18.

Rename the two recess features *CIRC* and *HEX*. These labels will make it easier to locate the features in the model tree. Save the part.

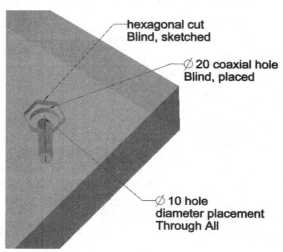

hexagonal cut
Blind, sketched

Ø 20 coaxial hole
Blind, placed

Ø 10 hole
diameter placement
Through All

Figure 18 Three features for UDF

Creating the UDF

In the Tools ribbon, select

> *UDF Library > Create*

Enter a name for the UDF as "**combo_hole**". Make it **Stand Alone** and include the reference part. **Add** the hole and the two recess features (pick them in the model tree, holding down CTRL), then select **Done** in the Select menu. Select the **Info** command (in the UDF FEATS menu) to open a window listing the features included in the UDF. Only the four desired features (including the make datum) should be listed. Close this window and select **Done/Return**.

You should now be in the **PROMPTS** menu. Read the message window and observe the highlighted axis (in blue) at the center of the block. Some of the references (like this datum axis) are used by more than one feature in the UDF. We can specify that Creo uses the same reference for all, or prompts for each feature individually. For the highlighted axis reference, select **Single | Done/Return** since we want to use the same axis reference for all features. For the prompt, enter "*axis for radial placement of hole*".

For the plane reference (FRONT), enter the prompt "*reference plane for hole placement angle*".

For the top surface reference, use **Single** and enter the prompt text "*hole placement surface*". These are all the reference prompts required. You can review these using **Next** in the MOD PRMPT menu and checking the message window. You can modify the prompt text at this time; otherwise, select **Done/Return**.

At this time, all the necessary UDF elements are defined. Let's exercise some of the options. We will make several dimensions available for the user to specify when the UDF is brought into the new part (make them *variable*): the radial placement dimension (ultimately to be the diameter of the UDF pattern circle), the placement angle (for the pattern leader), and the depth of the recess. The diameter of the hole and the size of the recess will be specified using a family table (*invariable*). The family table will also determine whether to use the circular or hexagonal recess.

Setting *Var Dims*

In the UDF elements window, select *Var Dims > Define*. All the feature dimensions appear. In the VAR DIMS menu select *Add > Select Dim*. The dimensions we want to make variable are (pick on each of these, the order doesn't matter; you don't need to use CTRL here, this is the old Creo interface)

> diameter of the placement circle of the hole (100)
> angle placement of the hole from FRONT datum (30°)
> depth of circular recess (10)
> depth of hexagonal recess (5)

After these have been selected, middle click and *Done/Return* in the VAR DIMS menu.

Now we will create text prompts for each of the variable dimensions. Creo will use a purple highlight to identify it in the graphics window. Enter the following prompt text for each highlighted dimension as it shows up (these may come up in a different order - pay attention to the purple highlighted dimension):

> *"hole placement angle"*
> *"hole pattern diameter"*
> *"depth (circ recess)"*
> *"depth (hex recess)"*

On to the next step...

Creating the Family Table

We will define a family table in the UDF that will specify the recess type (circular or hexagon), the hole diameter, and the recess size (diameter or width). In the UDF elements window, scroll down the elements list and select

> *Family Table > Define > Add Column > Feature*

Pick a surface on the circular recess or the name *CIRC* in the model tree. It will highlight in green. Then select a surface of the hexagon or pick on the name *HEX* in the model tree. It also highlights in green.

In the ADD ITEM area, select *Dimension*. Pick the dimension for the diameter of the center hole (currently 10) and enter *"hole_dia"* in the message window prompt. Similarly, pick the diameter dimension of the circular recess (currently 20) and enter *"CIRC_dia"* and the width across the hexagon (currently 25) and enter *"HEX_width"*. Accept the list of five family table items with *OK*.

In the FAMILY TABLE window, enter the data shown in Figure 19 to create the instances for the UDF. What does the symbol in the first column do? What do the "*" entries normally do? What is their effect in this family table? Note that in the generic part, both the circular and hexagonal features are present[3].

Type	Instance Name	F3 CIRC	F6 HEX	d0 HOLE_DIA	d3 CIRC_DIA	d6 HEX_WIDTH
	COMBO_HOLE	Y	Y	10.000000	20.000000	25.000000
🔒	C10-20	Y	N	10	20	*
🔒	C5-15	Y	N	5	15	*
🔒	H10-25	N	Y	10	*	25
🔒	H5-15	N	Y	5	*	15

Figure 19 Family table to define instances of the *combo_hole* UDF

Save the table and close the table editor with **OK**. This completes the UDF definition. In the elements window accept the definition with **OK**. Note that the message window indicates that *combo_hole* has been stored. Leave the UDF menu with **Done/Return**, then **Save** the original part and erase all objects from the session.

Using the UDF with Family Table

Create a new part **combo_test**. Set the units to millimeters. Create a base feature as shown in Figure 20. Create two datum axes through the centers of the cylindrical ends.

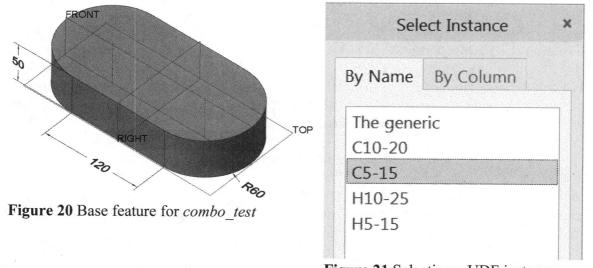

Figure 20 Base feature for *combo_test*

Figure 21 Selecting a UDF instance

---

[3] Note the differences from the regular Family Table editor: no **Preview** or **Verify** functions, no **Open** button. But there is an additional **Patternize** button. Check this out later.

Now we'll bring in the UDF and create an instance. In the **Model** ribbon, **Get Data** group:

User Defined Feature

Navigate to the working directory. You should see the UDFs *cover_pin* and *combo_hole*. Select the *combo_hole* and ***Open*** it. Since the UDF has a family table, the **Select Instance** window opens up as shown in Figure 21. Highlight the instance **C5-15** and ***Open*** it. Select the option to view the reference part (the source model) and click ***OK***.

The UDF Placement window now opens, this time with five tabs. In the **Placement** tab, we see (Figure 22) there are three references listed in the left pane. As each is selected, the feature will highlight in the reference model, and the appropriate prompt will be displayed. For the radial placement axis, pick the datum axis towards the right end of the part. For the angle reference, select FRONT. Finally, for the placement surface, pick the top of the part. All references are now defined.

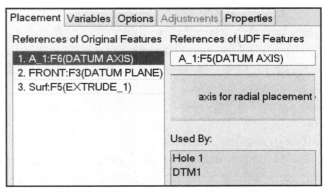

Figure 22 UDF Placement window - Placement

The UDF could now be placed. However, before we leave this window, select the **Variables** tab. See Figure 23. These are dimensions that were set to variable when the UDF was created. Select one of the rows in the table and observe the dimension prompt at the bottom of the window. Also, notice that the dimensions associated with the hexagon are not listed, since that feature is not included in the UDF at this point. Leave all the dimensions at their default values.

Owner	Source	Name	Value
Hole 1	Dimension	d7	100
Hole 1	Dimension	d8	30.0
CIRC	Dimension	d11	10

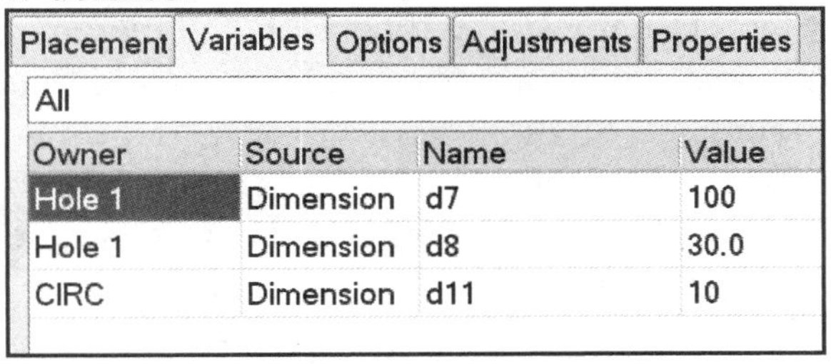

Figure 23 UDF variable dimensions

In the **Options** tab, set the dimensions to ***Lock***.

If you select the ***Adjustments*** tab, you can pick which side of the FRONT datum you want the hole on using the Flip button at the bottom. The location of the UDF we want is shown in Figure 24. Remember that the axis label appears at the "positive" end of the axis. If the axis is upside down, then the angle may be measured in the opposite direction from what you were expecting. Recall the orientation of the axis in the reference part and compare it to the orientation in the new model.

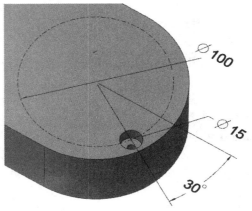

Figure 24 First UDF - group pattern leader

To correct this surprising placement, you could either enter a negative angle for the dimension in the *Variables* tab, or *Edit* the placement angle after the fact since it is unlocked. We are finished defining the UDF placement, so accept the feature.

Open the Model Tree to observe that the placed UDF appears as a group. Also, note that the feature HEX is not there (or even suppressed). We are going to pattern the group, so you might like to *Hide* the datum plane in the UDF group. Then, select the group in the model tree, and use the mini toolbar to select

Pattern

Select the angle dimension to increment in the first direction. Enter an increment of **90°** and specify **4** copies. There is nothing to do in the second direction. With the pattern completed, open the group pattern leader, select the hole, and *Edit* the pattern diameter dimension to **75** and the angle to **0°**. Try to change the diameter of the recessed hole. You can't, since it was invariable and set to *Lock*. What other dimensions are locked?

Bring in another instance of *combo_hole*, say the **H5-15** instance. Place it at **30°** from the FRONT datum using the axis at the other end of the new part. The placement diameter this time is **75** (change this in the **Variables** tab). In the **Options** tab, set the dimensions to *Lock*.

Finally, create a radial pattern of the hexagonal UDF. Use the angle dimension and increment by **60°**, with **6** copies. The final part should look like Figure 25.

See which dimension(s) you can modify on either of these patterns. Can you explain why?

Save the part and remove it from the session.

Figure 25 Final part

Exercise #3: Subordinate UDF

So far, the UDFs we have created and placed have been *Standalone* and *Independent*. When the UDF is in the new model, it has no connection to the original model. There may be cases where you want to maintain this connection. For example, consider the following scenario. Suppose you have a complicated group of features in the original model and these are used to create a UDF. The UDF is then used in several new part models. A design revision calls for the modification to one of the features contained in the UDF. You do not want to go through all the independent cases where the UDF was used. Instead, you would like to just change the original model and have the modifications show up in all the parts where the UDF has been used. This is what *Subordinate* and *UDF Driven* options can do for you. Not all changes in the original model can be accommodated. Basically, you can change the dimensions of the existing features. If you attempt to change (add or remove) the feature list or references, then Creo issues a warning. You might expect problems in any new models containing the UDF if drastic changes are made to its data structure.

Creating the Original Model

We'll demonstrate this using a very simple example. Create a new part called **udf_csinkhole_org** and set units to millimeters. Create the protrusion shown in Figure 26 and place a single datum point on the top center of the block. This must be an On Surface point that is placed on the desired surface with dimensions to appropriate references. Then, create a sketched hole using the **On Point** placement option. The sketch for the hole is shown in Figure 27. Place the hole on the datum point on the block.

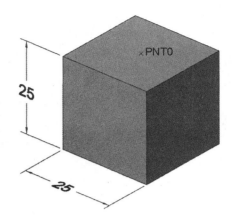

Figure 26 Base feature and placement point

The final original part model should appear as in Figure 28.

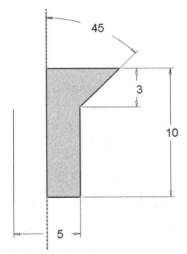

Figure 27 Dimensions for sketched hole

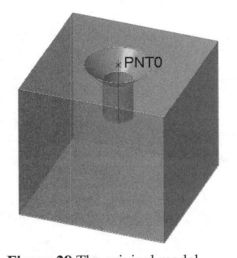

Figure 28 The original model

Creating the UDF

Now create the UDF as follows. In the **Tools** ribbon, select:

> ***UDF Library > Create***

Enter the name of the UDF as "**csink**". In the UDF OPTIONS menu

> ***Subordinate > Done***

This is the option that keeps the UDF tied to this original model. Click on the surface of the countersunk hole, then middle click. In the UDF FEATS menu, select ***Done/Return***.

The prompt for the datum point reference is "*hole placement point*". In the SET PROMPT menu select *Done/Return*.

Let's set the depth of the hole as variable. In the UDF elements window select

Var Dims > Define

and click on the depth dimension (10) then *Done/Return* in the VAR DIMS menu. Enter the prompt "*hole depth*".

We have finished defining the UDF. In the elements window select *OK > Done/Return*. The message window tells us that *csink* has been saved. Also, note that the part *udf_csinkhole_org* has been saved automatically. For *Subordinate* UDFs, the original model takes the place of the reference model (there is no part file *csink_gp.prt*) and is always saved with the UDF.

Erase all objects from the session.

Using the UDF

Start a new part called **csink_test**. Make a solid protrusion with a couple of datum points as shown in Figure 27.

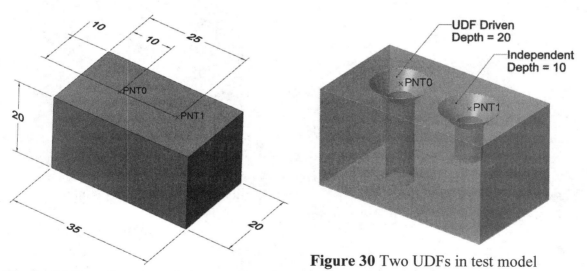

Figure 29 Base feature and datum points for test model

Figure 30 Two UDFs in test model

Now bring in the *csink* UDF. We will do this a couple of times to try out a couple of options. In the **Model** ribbon, select

User Defined Feature

Go to the working directory and open **csink.gph**. In the **Insert UDF** window that opens, check the option at the top, *Make features dependent on dimensions of UDF*. The other options are no longer valid. Select *OK*. The original model opens in a window, with the placement reference datum point highlighted in blue. Pick the datum point on the left end of the block. Select the **Variables** tab and change the hole depth to *20*. Finally, in the **Options tab**, note that the *Unlock* setting is no longer available, and *Lock* is the default option. Accept the UDF.

Bring in a second copy of **csink.gph**, leaving the default settings in the **Insert UDF** window. This will make the UDF independent. Note that the option to view the source model is checked automatically. The reference model will appear since the UDF is subordinate. Place the hole on the second datum point and set its depth to *10*. Notice that we can unlock the dimensions since we are creating an independent UDF.

The two *csink* holes are shown in Figure 30.

Exploring the Model

To demonstrate how the UDF is linked to the original model, open the original model file (it is in session at this time) *udf_csinkhole_org* and *Edit* the depth of the countersink from 3 to **5** (see Figure 31). *Regenerate* the part. Using the pull-down *Window* command, switch to the *csink_test* window and activate it.

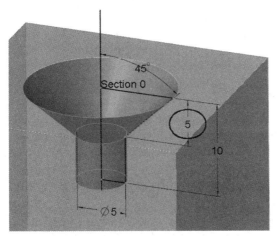

Figure 31 Modifying the depth of the countersink in the original model

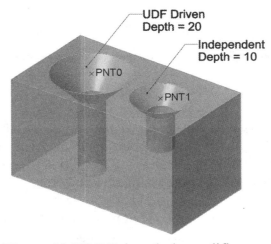

Figure 32 UDF Driven hole modifies with original model after *Update*

Nothing has changed yet in *csink_test*. We have to do two things for that to happen. In the **Model** ribbon, **Operations** group, select

> *UDF Operations > Update All*

Now *Regenerate* the model. Voilà! The UDF Driven *csink* on the left has been changed but not the one on the right, as shown in Figure 32.

To further demonstrate the link with the original model, save all your files and leave Creo. Go to your system and open your working directory. Notice the files *udf_csinkhole_org.prt* (the original model), *csink.gph* (the UDF definition), and *csink_test.prt* (our test part). Unlike *Standalone* UDFs there is no need for a reference part (like *csink_gp.gph*). **Delete** all but the highest numbered versions of these parts and then rename the original model file to something like *Xudf_csinkhole_org.prt*. Renaming it means that Creo won't be able to find it, and allows us to bring it back later.

Now bring up Creo and open the file *csink_test*. Both holes will appear, but the message window tells us there is a missing group. Try **Edit Definition** on the two holes. One hole will allow this but the other won't (without jumping through some hoops - see below!). Try to **Edit** the dimensions of the two holes. What happens and why? The moral is that if a UDF is created as *Subordinate* and then used as dependent or *UDF Driven*, don't lose the original model or you won't be able to do anything with the associated UDF (but see the next section for a remedy!).

Delete the *Independent* hole (on the right), and try to bring in the *csink* UDF again. You can't because the original model is not available (we renamed it).

Disassociating a UDF

Suppose you have brought in a dependent UDF and you want to break the connection to the original model. You can do this in the **Operations** group as follows:

> ### UDF Operations > Disassociate

and follow the prompts. If you do this, you are basically making the UDF *Independent*.

The Creo UDF Library

Earlier in this lesson we mentioned that there were a number of UDFs contained in the Creo UDF library. If installed, these can be found in the directory such as `ptc/objlib/featurelib`. To find the exact location on your installation you may have to consult your system administrator.

You can set up an option in your *config.pro* to automatically look in this UDF library when you launch **User Defined Feature**. The setting would be something like:

```
pro_group_dir c:/ptc/objlib/featurelib
```

The library contains dozens of UDFs for features commonly used in part modeling (cuts and protrusions of common sections like circles, hexagons, ellipses and so on), machine design, mold design, sheet metal, and piping design. Many of the UDFs produce ANSI standard geometries (for example, for threads) in both standard (i.e. inch) and metric forms. A directory tree for the feature library is shown in the table on the following page.

As of this writing, the Creo Help pages do not contain much information on the UDF library (such as how all the UDFs are named). Don't confuse this with the BASIC library of standard parts like fasteners.

Table 1: Directory Layout of the UDF Library

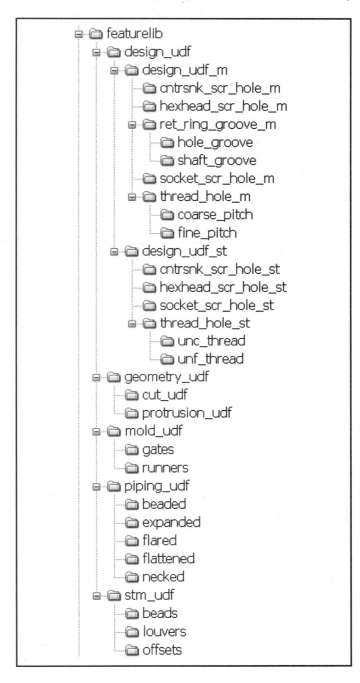

Conclusion

This lesson has introduced you to some methods to create and use User Defined Features. With careful planning and implementation, these can save you and others you work with a lot of time in model creation. They will also promote standardization among parts and assemblies where they are used.

We have, of necessity, only scratched the surface of using UDFs here. For example, we have not discussed creating or using UDFs in assemblies. Most of the steps and details are the same; however, there are some important differences. There are some features that cannot be included in a UDF used in an assembly (rounds, for instance). You are encouraged to explore the online help for further information.

In the next lesson we will look at some more tools for automating the geometry creation of a part using a program. We will also look at how layers can be used to simplify the on-screen display of parts and assemblies.

Questions for Review

1. What are the two primary motivations for using UDFs?
2. How many features can be contained in a UDF?
3. How many UDFs can be created from the original model?
4. What is the relation of the original model to the reference model?
5. Is a reference model always produced? Is it always required?
6. How does a UDF appear in the Model Tree?
7. What is the difference between a *Standalone* UDF and a *Subordinate* UDF?
8. Describe the purpose of the reference prompts.
9. What is the minimum set of information required to create a UDF?
10. For a standalone UDF called *trinket*, what are the complete names of the files stored for the UDF and its reference part?
11. What setting is used so that after the UDF is placed in a new model it no longer requires the UDF definition file?
12. What are the dimension settings available when a UDF is being placed in a new part? How do they differ in function?
13. Why would we like to define a UDF with as few references as possible?
14. What are *variable* and *invariable* dimensions? Where are these determined?
15. When you are creating features for a multi-feature UDF, what is the easiest way to make sure that optional features don't interfere with each other?
16. For a UDF involving a family table, is the table created before, during, or after we have identified the features involved in the UDF?
17. At what point in the creation of the family table can you specify names for the symbolic dimensions to appear as family table column headings?
18. What is the name of the UDF element that can be defined so that the user is prompted to enter values for dimensions when the UDF is being brought in to the new part?
19. How do you set up a UDF so that its geometry is always controlled by the original model? Why might you want to do this?
20. Is it possible to change the *invariable* dimensions if the UDF is **Independent**?
21. Is it possible to change the *invariable* dimensions if the UDF is a **UDF Driven**?
22. What does the **Update** command do? Does this command alone bring the new model up to the current geometry?
23. What happens when a part containing several **UDF Driven** groups is loaded?
24. What happens if the original model and UDF definition file are not in the current working directory (a) when the UDF is first added to a part, and (b) when the part is retrieved?
25. When a UDF has been used as **UDF Driven**, how do you break its association with the original model?
26. In what directory can you find the Creo Basic library UDFs on your system?
27. How do you set your *config.pro* so that the UDF feature library is automatically consulted when you want it?
28. Try out some of the library UDFs for cuts, protrusions, threads and others.
29. What happens if the units of the UDF and the receiving part are different?

Project Exercises

We'll create three parts this lesson. One will use two kinds of sweeps to let you review that material. The second is a pretty straightforward modeling exercise. The third will make use of a UDF that you will create.

The first part is the cart handle. The base feature is a simple constant section sweep. The handle is a variable section sweep using two elliptical sketches, one for the trajectory and one to define the radius of the section. You can invent your own dimensions for these.

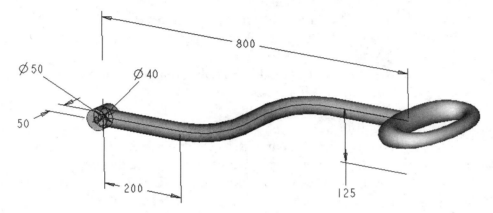

PART: *handle*

The second part is the main mounting plate for the wheel axle. Its construction should be straightforward. See the figures on the next page for more of the dimensions.

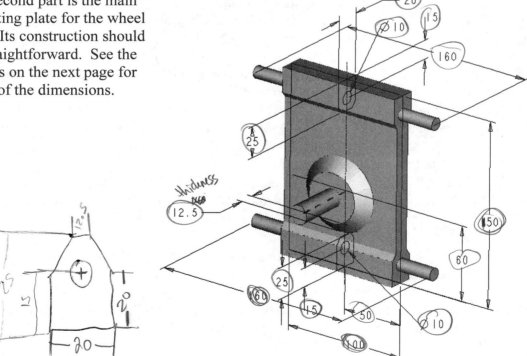

PART: *mount*

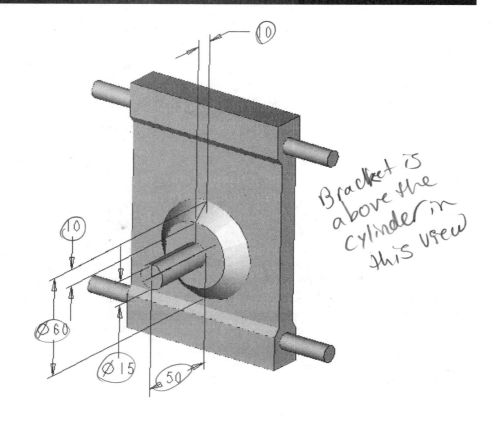

10

10

Ø60

Ø15

50

Bracket is above the cylinder in this view

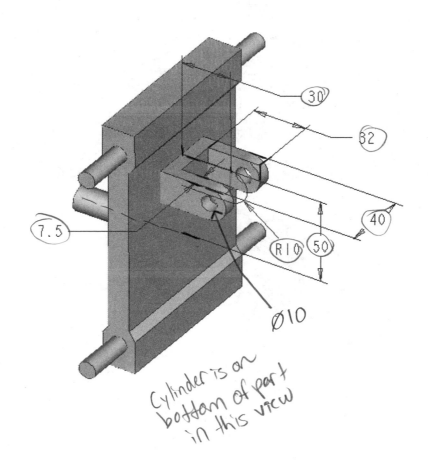

30

32

7.5

R10 50

40

Ø10

Cylinder is on bottom of part in this view

The final part is the wheel axle for the side wheels. This has four stud bolts arranged in a radial pattern. Create a UDF for this stud bolt using a revolved protrusion on a rotated *Make Datum*. Set up the UDF with a variable bolt length and radius to the center of the pattern. Then use the UDF in the creation of the part. The axle part has four studs that can be created with a radial pattern.

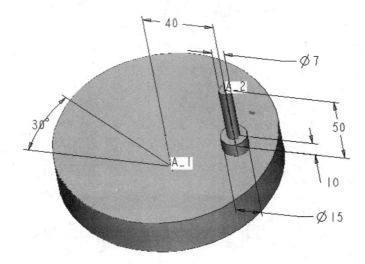

UDF: *stud*

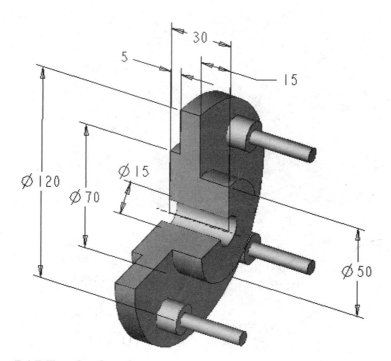

PART: *wheel_axle*

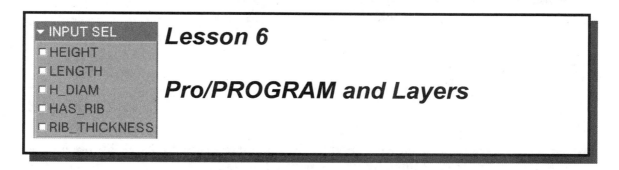

Synopsis

Using Pro/PROGRAM to create and run a part design file; input variables and conditionals; creating a family table using **Instantiate**; setting up and using Layers; default layers; adding items; controlling the layer display; layer settings in *config.pro*

Overview

This lesson is directed at a couple of very useful utility functions in Creo. The first of these, Pro/PROGRAM, allows you to create a script to control what happens during the regeneration of a part. This includes prompting the user for variables that can be used to assign dimensions, or to control the execution of the script. Pro/PROGRAM offers a convenient method to create or add new instances to family tables. The second major utility function concerns the topic of layers. These are used to organize features into logical groupings. Layers can be used to control the display or the selection process for other feature manipulations such as suppression. The assignment of features to layers can be done manually or automatically using settings created only for a single session or universally using entries in the configuration file *config.pro*.

Each of these utility functions will be demonstrated and explored using fairly simple parts.

Pro/PROGRAM

Pro/PROGRAM is a utility that lets you specify input variables (used for dimensions), set up relations, and modify or control the regeneration process for parts and assemblies. PROGRAM is based on a textual description of the part or assembly called the *program* or *design*. Using syntax and logical structure similar to a programming language, you can specify names and values of variables, perform arithmetic operations and logical branching, and other functions that will affect the regeneration sequence. The design file can be as simple or elaborate as you like. PROGRAM lets you and users of your models very quickly create alternative part models. It is very easy for the user to set values for variables and determine whether or not to include features in parts or components in assemblies. The model can also react to these changes by substituting alternate features or components. PROGRAM is also a very convenient way to set up a family table. Its main advantage is that all this can be done without requiring the user to have in-depth and comprehensive knowledge of the model.

PROGRAM Elements

Every part or assembly has a default program or design embedded in it that is created automatically. A very simple part, such as the cylinder in Figure 1, illustrates the basic structure of this design file, shown in Figure 2. Note that most sections of this default program are empty. In the following, this listing will be referred to as the *program* or the *design*. At times, the listing will be saved to disk, where it becomes the *design file*.

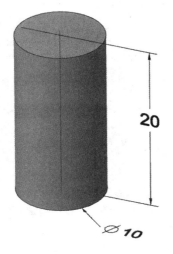

Figure 1 A single feature part

```
VERSION 5.0
REVNUM 103
LISTING FOR PART SIMPCYL

INPUT
END INPUT

RELATIONS
END RELATIONS

  ADD FEATURE (initial number 1)
  INTERNAL FEATURE ID   31
                :
  END ADD

MASSPROP
END MASSPROP
```

Figure 2 The (abbreviated) design file for the simple cylinder of Figure 1

The program or design contains five areas. At the top are the software version and the model's name. Then come two areas (initially empty as shown in Figure 2) where we can define program *Inputs* and part *Relations*. The biggest area of the file contains sets of *ADD ... END ADD* statements, each set enclosing a single feature. In an assembly design this area is used to *ADD* components to the assembly. At the end is a section where mass properties functions can be placed.

Syntax of PROGRAM Statements

In the space available here, we cannot go into all the variations of syntax or built-in functions available with Pro/PROGRAM. All of this material can be found in the on-line help[1]. There is enough here, however, to be quite useful and to give you a good idea of what PROGRAM can do. We will go deep enough into its operation so that you can implement more advanced functions described in the on-line help on your own.

---

[1] The on-line help for Pro/PROGRAM can be found by selecting **Fundamentals** then pick the topic **Program** and **Editing a Design**.

The *INPUT* Syntax

The INPUT section of the design file is where we indicate the names and types of variables that will be used in the design, part, or assembly. Variable names are alphanumeric and must begin with a letter. There are three types of variables allowed: *Number*, *Yes_No*, and *String*. Along with each variable we can specify a text prompt string that will appear in the message window at the appropriate time (when PROGRAM requires input). Some examples of the format of the input definitions are shown below:

```
INPUT

        plate_thickness number
        "Enter the plate thickness (mm)"

        include_flange Yes_No
        "Do you want to include the flange in the part?"

        material string
        "Enter the part material (ABS, PVC, or POLY)"

END INPUT
```

Note that the prompts occur immediately after the variable name and type are defined and are enclosed in double quotes. Although the text is shown here in lower case, all characters except those included between double quotes will be converted to upper case. String variables can be names of part or assembly files. In some cases, when used as a string variable, names of parts and/or assemblies must have the *prt* or *asm* extensions.

The *RELATIONS* Syntax

The RELATIONS section of the design contains any relations defined for the part or assembly. As such, the syntax of relations is the same as in other areas of Creo. They can be created with the ***Relations*** command or within the Pro/PROGRAM editor.

The *IF...ENDIF* Syntax

Pro/PROGRAM also allows logical branching using boolean variables. This takes the form

```
IF {boolean expression}
        {do this}            ← if boolean expression is TRUE
ELSE
        {do that}            ← if boolean expression is FALSE
ENDIF
```

The boolean expression can involve any of the input variables, part parameters or dimensions, or string variables. This form of branching can occur almost anywhere within the design file. IF statements can also be nested to provide very deep logical branching.

Some examples of the syntax are as follows (required syntax elements in upper case):

```
IF num_holes > 10
    hole_diam = 12
    hole_space = 60
ENDIF

IF material == "steel"
    wall_thickness = 5.0
ELSE
    wall_thickness = 8.0
ENDIF
```

In the second example, note the difference between the "==" sign used in the boolean expression to test for equality and the "=" sign used for assignment of a value. We will see a couple of examples of this later in this lesson.

Pro/PROGRAM contains several other syntax elements and built-in functions. For example, it is possible in an assembly to ADD an instance from a part family table by searching for an instance with prescribed properties. See the on-line documentation for detailed discussion of these functions.

Example: A simple bracket with optional rib

We will create a simple part to demonstrate the basic principles of PROGRAM. We will control the regeneration of a part using the syntax elements presented above. The design will include several *Input* variables and an optional feature (a rib). We will see how to automatically create a Family Table for the part using the *Instantiate* command.

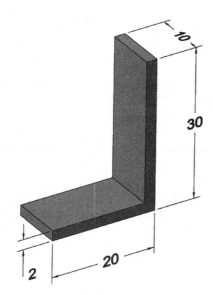

Figure 3 *Bracket* base feature

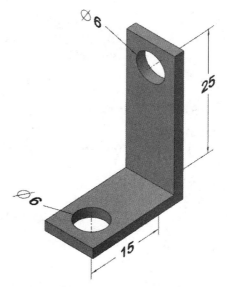

Figure 4 Adding holes to base feature

Creating the Part

Start a new part called **bracket**. Create an L-shaped base feature as a both-sides solid protrusion off the RIGHT datum. The bracket rests on the TOP datum and the dimensions are shown in Figure 3. Add the two holes (linear placement, one sided, *Thru All*) shown in Figure 4, noting that the holes are aligned with the RIGHT datum on the center plane of the part. Round off the ends of the legs of the bracket using cuts (Figure 5). Each cut is coaxial with the holes.

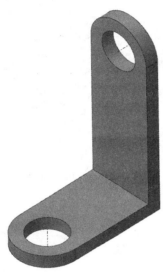

Figure 5 Rounded ends using *Cut*

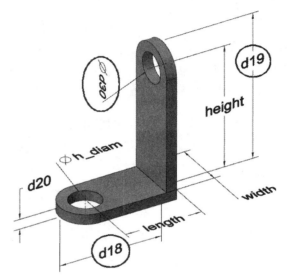

Figure 6 Dimension symbols renamed and some to note

To prepare for using PROGRAM, we will rename some of the symbolic dimensions. Click on each feature to display the dimensions. Select the following dimensions then change the names using the dashboard to the following:

> **height** - the vertical height to the hole on the back leg (25)
> **length** - the length along the base leg to the front hole (15)
> **width** - the width of the bracket (10)
> **h_diam** - the hole diameter (6).

Also, note your symbolic dimension names for the circled dimensions d18, d19, and d30 shown in Figure 6 - we'll need these below.

We are going to set up some relations for this part. These will work together with the part program (actually, they are included in it) to determine the geometry during regeneration. We want to set the width of the bracket and the length of each leg based on the hole locations and to maintain a clearance of 2 around the holes. In the **Model Intent** group, select

> ***Relations***

and enter the following (Note that your dimension symbols *dxx* may be slightly different from those shown in Figure 6):

```
/* top hole diameter
d16 = h_diam
/* height of back leg maintains clearance around hole
d11 = height + h_diam / 2 + 2
/* length of front leg maintains clearance around hole
d12 = length + h_diam / 2 + 2
/* width of bracket to maintain clearance
width = h_diam + 4
```

As usual, after you have set up some relations, check them out to make sure they are working correctly. Try to modify the vertical leg height - you can't since it is driven by a relation. Try other dimensions to see if they can be modified. Do both hole diameters become equal? Is our intent of maintaining the clearance around the hole implemented properly? If you find any errors in the dimensions, fix them now using **Relations**. Return all values to the starting values when you are finished (height = 25, length = 15, hole diameter = 6).

We are now ready to set up the part program. Save the part before we start experimenting!

Creating *Input* Variables

In the **Model Intent** group, select

> *Program*

The program menu comes up to give you options for dealing with the current design. To see the format of the "default" design (that is, the unmodified part program), select

> *Show Design*

An information window will open containing all the sections of the part program. The major sections were described above (*Input*, *Relations*, *Features*, *Mass Properties*). Notice that the relations we created above are listed in the design. Close this window and in the PROGRAM menu, select

> *Edit Design*

Depending on how your system is set up, this will bring up either the Pro/TABLE editor or a system text editor. Notice that the design file name is *bracket.pls*. In an assembly, the design file extension is *als*.

Add the lines shown in Figure 7 between the INPUT ... END INPUT lines to declare the names, types, and prompts for the input variables. We recognize these as the symbolic names for the bracket dimensions (*height, length, h_diam*). If you enter characters in lower case, they will automatically be converted to upper case when the program is saved.

```
LISTING FOR PART BRACKET

INPUT
 HEIGHT NUMBER
 "Enter height of back leg to hole:"
 LENGTH NUMBER
 "Enter the length of base leg to hole:"
 H_DIAM NUMBER
 "Enter the hole diameter:"
END INPUT
```

Figure 7 Creating input variables

Incorporating the Design

Exit the editor and save the file on the way out. When you leave the design text editor, you are asked if you want to *incorporate* the changed program in the model. Here is what happens next:

- If you select *Yes*: The program is added to or embedded in the model, the part geometry will be updated (regenerated), and the *pls* file will be removed from your working directory. The next time you call up the ***Edit Design*** command, there is only one source for the design file, that is, the one embedded in the model.

- If you select *No*, then the new design is NOT embedded in the model (and thus not changing the current geometry). Furthermore, the *pls* file will remain on the hard disk. The next time you call up the ***Edit Design*** command, you would be prompted to select which design file to work on: the one contained within the model (***From Model***) or the one on the disk (***From File***). These could be different. There are several reasons why you might have a different copy of the design on the disk without incorporating it: you may have been editing the design file and realized you had to quit the editor and make other modifications to the part in order to make the design work when it is incorporated, or you may be trying out several different ideas in different design files. In the latter case, you will have to be careful about naming the "active" design file; it must always be of the form *<partname>.pls*. If multiple *pls* files exist for the same part, they will be incrementally numbered. ***From File*** reads the highest numbered file. Also, remember that once the design is incorporated into the model, the *pls* file is deleted.

For now, select *Yes*.

The GET INPUT window opens, a signal that the program is running (see the message window). The first thing that happens is the processing of the INPUT statements. Creo wants to know where to obtain values for these variables. Select ***Current Vals***.

Now select ***Show Design***. This time, the information window also shows us the current values assigned to the variables in the INPUT section. Close the information window and select ***Done/Return***.

We now have a program defined for the part.

Running the Program

The program will execute whenever the part is regenerated (including when it is first retrieved). Do that now:

Regenerate (or press CTRL-G)

The GET INPUT window opens again. This time, select

Enter

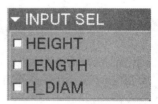

A small menu (INPUT SEL) appears as shown in Figure 8 listing the variables defined in the INPUT section of the program. You can check any or all of the boxes, then *Done Sel*. You will be prompted in the message window to enter new values for the selected variables. Note the current value is always the default and can be accepted by just pressing the Enter key (or middle mouse click). When all selected variables have been set, the part will regenerate. Try this out for different combinations of variables by selecting

Figure 8 Selecting *Input* variables

Regenerate over again each time. Before you proceed, return the values to the original ones (height = 25, length = 15, hole diameter = 6) and save the part.

Modifying the Part

Add the *Profile Rib* feature shown in Figure 9 to the part. This is sketched on the RIGHT datum plane. The top and front edge of the rib are 1 unit from the edge of the holes. The rib thickness (symmetric about RIGHT) is also 1.

When the rib has been created, rename the thickness dimension symbol to *rib_thickness*.

Applying Conditionals

Conditionals (IF statements) can be added at many places in the design file to control the regeneration of the part. In the bracket, we will set up a variable to allow for the optional creation of the rib feature. If the rib is to be included, we will prompt for its thickness. Select (in the **Model Intent** group)

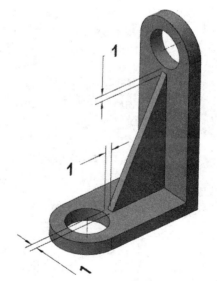

Figure 9 Adding the optional rib

Program > Edit Design

If you have been following the sequence here precisely, then the design editor should come up immediately. If you have to make a choice, use *From Model*.

In the INPUT area of the design file, add the variables *has_rib* (of type *yes_no*) and *rib_thickness* (of type *number*). See the listing in Figure 10. Each variable has a prompt. Note that in the INPUT section, if the user sets the value of *has_rib* to **YES**, then Creo will prompt for the value for the rib thickness. To control the regeneration of the rib, add the IF..ENDIF statements to the design file to bracket the ADD..END ADD statements for the rib feature (near the end of the design file) as shown in Figure 11. Thus, if we set the variable *has_rib* to **NO**, then the ADD FEATURE statements for the rib will be skipped. The rib is not deleted from the part, just suppressed.

```
INPUT
 HEIGHT NUMBER
 "Enter height of back leg to hole:"
 LENGTH NUMBER
 "Enter length of base leg to hole:"
 H_DIAM NUMBER
 "Enter hole diameter:"

has_rib yes_no
"Does the bracket have a rib?"
if has_rib == yes
    rib_thickness number
    "Enter thickness of the rib:"
endif

END INPUT
```

Figure 10 Program with conditionals (partial listing 1 of 2) to control input of variables

```
if has_rib == yes
    ADD FEATURE 23
    INTERNAL FEATURE ID   555
    PARENTS = 345(#18) 398(#19) 425(#20) 1(#1) 3(#2)

PROFILE RIB: Profile Rib

   NO.          ELEMENT NAME        INFO
   ---          ------------        ----
    1           Feature Name        Defined
    2           Sketch compound     Defined
    .
    3           Rib Thickness       2
    4           Side Options        Symmetrical

       SECTION NAME = Section 1
       OPEN SECTION


       FEATURE'S DIMENSIONS:
       rib_thickness = (Displayed:) 1
                     (    Stored:) 1.0 ( 0.0, -0.0 )
       d43 = (Displayed:) 1
             (    Stored:) 1.0 ( 0.0, -0.0 )
       d44 = (Displayed:) 1
             (    Stored:) 1.0 ( 0.0, -0.0 )
       END ADD
endif
```

Figure 11 Program with conditionals (partial listing 2 of 2) to control regeneration of feature

To see how Creo responds to a program error, leave out the ENDIF shown in Figure 10. Save the file and exit the text editor. The error is immediately detected, and you can either abort the program change or select *Edit* to re-enter the program editor. Do that now and you should find that an error message

```
!*** ERR: file contains more IF's than ENDIF's
```

is placed at the end of the INPUT section. This is close enough for us to locate the error. Fix it now, save the file, and exit the editor.

Incorporate these changes into the model. When the GET INPUT menu opens, select *Enter*. The new INPUT SEL menu opens (Figure 12) showing the new variables we have added. Check the *has_rib* variable, then *Done Sel*. Set the rib to **NO**. The prompt for rib thickness will be skipped and the part should regenerate without the rib. Open the model tree to see the suppressed feature. Can it be resumed using the command in the mini toolbar?

Figure 12 Selecting *Input* variables for Regeneration

Select *Regenerate* and check the setting for *has_rib*. This time the rib appears but has the previous thickness. Select *Regenerate* again and change both variables.

This illustrates the basic operation of the program and how the user can easily modify values in the design.

Try this: double-click on the rib, and enter a new value for the rib thickness (say **2**) by clicking on its displayed dimension. Now select **Regenerate**. Once again you see the GET INPUT menu. Choose **Current Vals**. The part regenerates with the new value for rib thickness. Thus, there are two ways to change the values of variables contained in the INPUT section of the design file; that is, using

> **Regenerate > Enter**
>
> or by   **Edit Dimensions > Regenerate > Current Vals**

Instantiating to a Family Table

A useful function available in conjunction with PROGRAM is the automatic creation of a family table for the part. A new instance using the current geometry is created if we go to the **Model Intent** group and select (do this now)

> ### *Program > Instantiate*

Enter the name "bracket". If this is the first instance, the family table will be automatically created, using the INPUT variables as column headings. The current part becomes the first row (and hence the generic) in the table. If there are previous instances, a new entry is made to the family table.

Create another geometry by selecting **Regenerate** and setting the variable values shown in the second line in Figure 13. Then use **Program > Instantiate** again, entering the name given in the table (**B3020-6-R1**) as shown in Figure 13.

Instance Name	Common Name	d33 HEIGHT	d29 LENGTH	d26 H_DIAM	HAS_RIB	d42 RIB_THICKNESS
BRACKET	bracket.prt	10.0	8.0	3.0	NO	1.0
B3020-6-R1	B3020-6-R1	30.0	20.0	6.0	YES	1.0
B1008-3	B1008-3	10.0	8.0	3.0	NO	1.0

Figure 13 Family table created with Pro/PROGRAM

Regenerate the part using the values shown in the third line, and instantiate the part.

Now open up the family table for the part using (in the **Model Intent** group)

> ### *Family Table*

Notice that the family table automatically contains our symbolic names in the column headings. The generic part (**bracket**, in this case) always contains the values of the variables present in the model when it was most recently instantiated by Pro/PROGRAM. The use of the "*" symbol in this table is therefore quite dangerous, and Creo will not use one if it generates the instance for you. Create some more instances for this family table.

Then, try to modify any of the values driven by the family table. There are several ways to do this:

> **Edit Dimensions** > (change dimension values) > **Regenerate**
> **> Current Vals > Program > Instantiate**

OR

> **Regenerate > Enter > Program > Instantiate**

OR

> **Family Tab** {edit one or more cells}

What happens to the part and the family table in each case? Do the instances automatically appear as new parts on disk?

Reading Values from a File

Another useful function to investigate is reading variables from a file. For example, using your system text editor, create a file **bracket.txt** containing the following:

```
h_diam = 5
length = 15
height = 30
has_rib = yes
rib_thickness = 2
```

Save this as a simple text file (ASCII) in the current working directory. Then in Creo, select

> **Regenerate > Read File > [bracket.txt]**

Now use **Program > Instantiate** to add this to the family table (call it **B3015-5-R2**). Pretty slick! The interpretation of this text file is as follows:

- variable names in the file that are not used are ignored (for example, if we had set **has_rib** to **no** we would not need **rib_thickness**)
- variable names required by PROGRAM but not included in the file use the current model values
- variable names that are unrecognized, misspelled (or have a typo) are ignored

We have finished with the bracket part, so save it and erase it from your session.

Where to go from here?

This concludes our quick introduction to Pro/PROGRAM and hopefully you have caught an idea of how it might be used. There is lots more you can do with this function. For example, you can select UDFs conditionally, or substitute different UDFs depending on variable values. In assembly mode, you can control the presence or absence of components, interchange components or family table instances, pass variables from an assembly down to its sub-components and control execution of programs there. See the on-line documentation for further details.

Layers

The concept of layers has been around in CAD packages for many years. In 2D drawing packages, layers are used to organize related entities on a separate transparent "sheet" of the drawing. The sheets are stacked in "layers" to produce the drawing. For example, a building layout might have layers corresponding to the structural details, piping, air conditioning, and electrical systems. Each layer typically has independent view control so that, for example, the plumber doesn't have to use a drawing covered with details of the electrical system, whose layer can be turned off.

Layers in a Creo model (part or assembly) perform a similar function, although of course we are not dealing with 2D entities. Layers in Creo contain features in part mode, or components and assembly features in assembly mode. The primary purpose of layers is to organize the feature/component database to simplify the operation of the program. A side benefit is that layers will speed up the graphics display by helping to remove undesired or unnecessary display elements. In operation, layers are used for the following:

- **Controlling the display of some features and components**. In part mode, this affects only non-geometric features on the layer (see discussion later on geometric and non-geometric features). In assembly mode, display control affects both features and components. Independent control over the display state of each layer is possible. The states are *Hide* and *Isolate*. A layer that is set to *Hide* will not display, although the features still exist in the model (they are not suppressed). *Isolate* is the inverse of *Hide* - all layers except the one set to *Isolate* are hidden. This is useful in an assembly when you want only components on a particular layer to be displayed.

- **Organizing features/components into logical or related groups**. Many operations that require feature/component selection, such as suppressing, can be carried out simultaneously on all members of the selected layer(s). For example, a layer might contain all the rounds in the model, or a set of diverse features all relating to one area of a part. This level of organization can be very convenient if planned properly.

We will demonstrate most of the functionality of layers by creating a simple part and then experimenting with the use of layers.

Creating The Model

The model we are going to create is shown in Figure 14. This is a simplified half model of a human mouth and throat (the esophagus) that has been used to study airflow and the delivery of aerosol medications. Start a new part called **throat** and set units to millimeters. The base feature is a simple protrusion on the TOP datum with the dimensions shown in Figure 15. We'll take this opportunity to review a couple of features we have seen before - a variable section sweep and a pattern table.

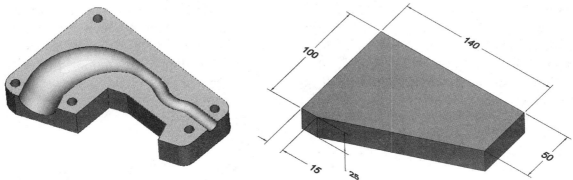

Figure 14 Finished *throat* model   **Figure 15** Base feature of *throat* model

The throat passage will be created using a variable section swept cut. To define the origin trajectory of the sweep, sketch a curve on the top of the block as shown in Figure 16 (constraint display has been turned off). A second trajectory (Figure 17) will be used to provide a reference that will define the radius of the cut. Note that several segments near the right end of the second datum curve were defined with the **Offset Edge** option in Sketcher.

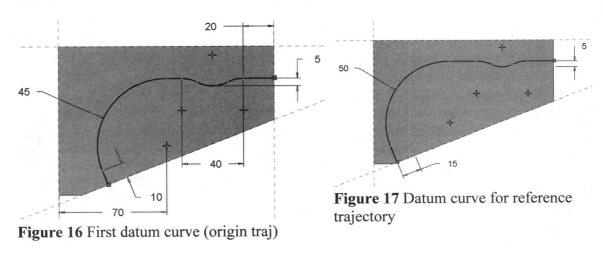

Figure 16 First datum curve (origin traj)

Figure 17 Datum curve for reference trajectory

Now use the datum curves to define a variable section sweep. The first datum curve will be the origin trajectory down the center of the sweep. The second trajectory will be used to define the radius of the cut. See Figure 18. To refresh your memory of the variable section sweep, here are the commands: Select **Sweep** then pick on the first datum curve (it will become the origin trajectory). With CTRL, pick on the second curve (it becomes Chain 1). Now select

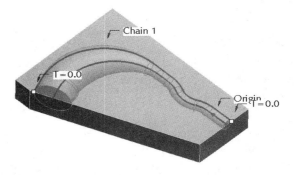

Figure 18 Variable section sweep cut

Sketch and draw a circle centered on the origin of the cross hair and snapping to the reference from Chain 1. Accept the sketch. Make sure you create a **Solid** and select the **Remove** button. If all is well, accept it.

Now we'll put some holes in the base block. Start by making the first hole in the rear left corner. This uses a linear placement, with dimensions to the left and back edges (both 10mm). The ***Through All*** hole diameter is 7mm. This will be used as a pattern leader for a table driven pattern. Prior to making the table, modify the dimension symbols of the pattern leader as shown in Figure 19. This makes the table easier to read. Then create the pattern table shown in Figure 20.

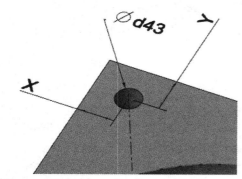

Figure 19 Pattern leader for hole

!	Table name HOLE_PATT.	
!		
! idx	X(10.0)	Y(10.0)
1	130.0	10.0
2	130.0	36.0
3	8.0	92.0
4	50.0	76.0

Figure 20 Pattern table for holes

Finish off the model to this point by adding some R10 rounds to four of the vertical edges of the block as in Figure 21. Save the part.

We are now ready to start playing with the layers.

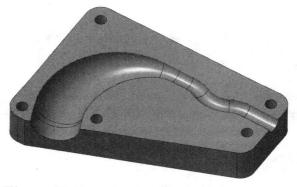

Figure 21 Rounds added to model

Using Layers

Layer operations are performed using the Layer Tree display in the Navigator window. To view the Layer Tree, open the Navigator window and select the model tree tab. Then either select ***Show > Layer Tree*** or pick the ***Layers*** button in the **View** ribbon. This button is a toggle.

If your system has any default layers defined (for example in your default part template), they will be shown in the Navigator. We will assume that you have no layers defined. If there are any layers already present, select them all and in the RMB menu select ***Delete Layer***. This just deletes the layers, not the features stored on them.

There are two methods that can be employed within this window to launch the layer-related commands:

- using the Layer pull-down menu in the Navigator
- right-clicking within the layer tree window

Let's see how these work...

Creating Layers

First we need to create some layers to put the features on. In the Navigator menu, select

Layer > New Layer

or use the RMB pop-up menu (cursor must be within the Navigator). This opens the *Layer Properties* dialog window. This is the window used to name layers and specify which items are to be placed on each layer.

We will create five layers first, then discuss other functions in the window. Type in each of the following into the **Name** field. When you press the *Enter* key after each, the new layer will be added to the layer tree. You can then select *New Layer* again and type in the next layer name. Add these layers:

```
my_datum_curves
my_protrusions
my_rounds
my_holes
my_cuts
```

The "*my_*" prefix will help us to identify these layer names and not get confused with feature types or default layers.

If you make a mistake typing, or create a layer you don't want, the offending layer can be renamed or deleted by selecting it in the Layer Tree window and holding down the RMB. This brings up the pop-up menu shown in Figure 22. Select the appropriate command to either *Delete* the layer, *Rename* it, or access its *Layer Properties*.

Adding Items to a Layer (with *Include*)

We will look at a number of options for adding items to the layers. To start, select the layer **my_datum_curves** in the layer tree, then with the RMB pop-up, select *Layer Properties*. The *Contents* and *Rules* tabs in the **Layer Properties** dialog are the two methods used to add items to layers. The same item(s) can be added to several layers at the same time, and items added to any particular layer can come from a combination of both methods of selection.

Activate
Deactivate

New Layer...
Copy Layer
Paste Layer
Delete Layer
Rename
Layer Properties...

Cut Item
Copy Item
Paste Item

Figure 22 Part of the RMB pop-up menu in Layer window

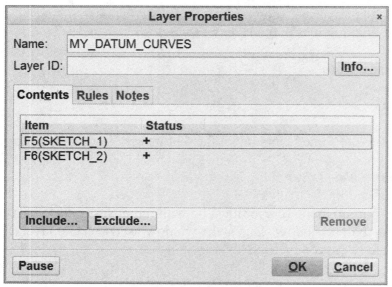

Figure 23 Layer Properties dialog window

The **Contents** tab opens a collector into which we can manually place model features. Note the ***Include*** button below the collector. When this button is selected you can pick on any feature in either the graphics window or the model tree and it will be added to the layer. To open the model tree, select ***Show > Model Tree*** in the Navigator. The **Layer Properties** window will stay open when the model tree is being displayed. Pick the two datum curves used to define the sweep (preselection highlighting may come in handy here). The **Layer Properties** window should look like Figure 23 above. If you accidentally add the wrong feature to the collector, just highlight it and select the ***Remove*** button at the lower right. Select ***OK*** in the **Layer Properties** window. Show the layer tree in the Navigator.

The layer **my_datum_curves** can be expanded to list the two curves. If these are picked, they will highlight in the graphics window in green. If the layer name is selected, all items in that layer highlight in the graphics window.

Select the layer **my_holes** and open the **Layer Properties** window. With ***Include*** selected, go to the model tree and pick on the pattern feature. Select ***OK*** in the **Layer Properties** window. Back in the layer tree, the pattern is listed and not the individual holes. We will use a different selection method below to list them individually.

Moving Items between Layers (with *Cut / Copy / Paste*)

In the layer tree, select the **my_protrusions** layer and open the **Layer Properties** window. Let's make a "mistake" here. With the ***Include*** button pressed, pick on one of the rounded corners. This puts the feature into the protrusions layer. This feature should probably be on a different layer. Close the **Layer Properties** window and expand the **my_protrusions** layer in the layer tree. Select the feature listed there. Now you can use the ***Cut Item*** (or ***Copy Item***) commands in the RMB pop-up menu - these operate the same as regular Windows commands. Select the layer where you want the feature (**my_rounds**), then in the RMB pop-up menu select ***Paste Item***.

To cut or copy all items on a layer, select the layer name and in the RMB pop-up menu, select the command *Select Items*. Try that with the **my_datum_curves** layer. Once they are selected, you still need to use *Copy Item*. Copy the two features listed there to the **my_cuts** layer. What happens if you try to select a layer and copy it to another layer? The result will be nested layers. Try it!

WARNING: If you select an item in a layer and press the *Delete* key, you are asking to delete the feature from the model. Fortunately, you are asked to confirm this and the default response is *Cancel*. If you just want to take the feature off the layer, use the *Remove* command either in the RMB pop-up menu or in the Layer Properties collector.

Adding Items to a Layer (with *Rules*)

Rules can be set up to automatically determine what will be included on a chosen layer. Rules can act retroactively (i.e. examining the entire model created prior to when the rule is added) or only after the rule is implemented. Let's create a rule so that all protrusions will automatically be placed in the **my_protrusions** layer.

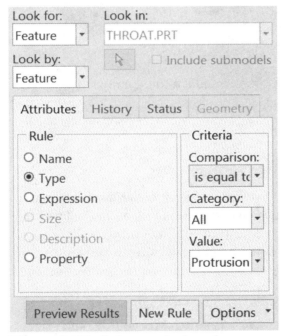

Figure 24 Defining a rule for a layer

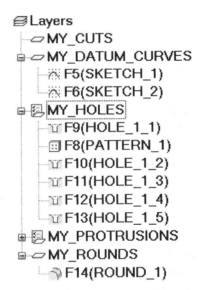

Figure 25 Layer tree with features added by rules

Open the *Layer Properties* window for the **my_protrusions** layer. Select the *Rules* tab above the collector. Pick the *Options* button, and make sure that *Independent* is checked (read the pop-up tip). Now, the *Edit Rules* button below the collector is active, so pick it. This opens a new window where we can specify selection rules for the layer (see Figure 24). These rules can be based on **Attributes** (for example: feature names and types), model **History** (for example: when created), or feature **Status** (for example: suppressed features). The **Look For** selection at the top of the window should be *Feature*. Pick the **Attributes** tab, select the *Type* radio button, then in the pull-down lists, set *Comparison(is equal to)* and *Value(Protrusion)*. Select *Preview Results*. A collector lists the entities (only one in this case) that satisfy the rule. Select *OK* to add this rule to the Layer Properties rule collector. Select the *Options* button again, and pick *Associative* (read the pop-up tip). Note that multiple rules can be defined for a particular layer. Accept the **Layer Properties** dialog. The protrusion feature is added to the **my_protrusions** layer. Also, the icon associated with this layer in the layer tree has changed to indicate that it is using rule-based selection.

Select the **my_holes** layer (currently contains the pattern), and add a rule to select all hole features. Don't forget to make it *Associative*. Accept the new layer property. Back in the layer tree, the five individual holes are listed in the layer. See Figure 25.

Add a rule to the **my_cuts** layer properties so that all cuts will be automatically added to this layer. The variable section sweep should now be added. To try out the rule, create the cut shown in Figure 26. Open up the layer tree to confirm that the **my_cuts** layer now has a new feature.

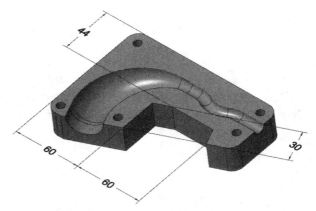

Figure 26 A *cut* added to the model

So far, we have discussed how to put features onto layers, but not what to do with the layers themselves. As mentioned above, there are two basic reasons to organize the model in layers: display control and selection sets.

Controlling Layer Display (with *Hide / Unhide*)

Select the **my_datum_curves** layer. In the RMB pop-up menu, select *Hide*. When you *Repaint* the screen, the datum curves are no longer visible. Also, the icon in the layer tree changes to indicate that entities on the layer are hidden (the option *Settings > Hidden Layers* must be checked). Features on a hidden layer are still regenerated and are part of the model. The situation is similar to turning the display of datum planes or axes on and off without affecting their children.

To get the feature display back, select the hidden layer, use the RMB pop-up, and select *Show*.

The *Hide* command has three important characteristics:

1.   The *Hide* command (in part mode) will only affect some features. Basically, these are features that do not directly affect the geometry that would change the solid-ness of the part. These are sometimes called *nogeometry* features (datum planes, curves, points, and others). An easy way to remember these is that they do not directly affect mass properties of the model.

2.   A layer can contain both geometry (i.e. solid) and nogeometry features. Although you can set a layer containing solid features to *Hide*, only the nogeometry features are affected.

3.   If a nogeometry feature is present on a number of layers, any one of them can be used to hide the feature.

For examples of these, try hiding various layers in the model we have made so far. For example, copy the sketched curves to the **my_rounds** layer. *Unhide* the **my_datum_curves** layer and then *Hide* the **my_rounds** layer. What actually gets hidden?

Feature Selection using Layers

An important function of layers in part mode (probably more than Hiding) is the ability to identify and select sets of similar features. For example, suppose you want to suppress all the rounds in a part. Open the layer tree; highlight the **my_rounds** layer. In the RMB pop-up menu, choose *Select Items* to pick everything on the layer. Move your cursor to the graphics window, hold down the RMB and select *Suppress* in the mini toolbar. Expand the **my_rounds** layer. A small symbol ■ indicates the suppressed feature. Since a layer can contain numerous common items of different types, this way of selecting features makes it convenient to pick features that could be stored all over the model and model tree. You can also resume features by layer.

Another possibility that you could try for a very complicated part is to place all related features on the same layer. If you are working on another region of the part, you could select the layer and suppress all the unnecessary features simultaneously to get them out of the way. Since items can appear on more than one layer, this gives you a lot of flexibility in organizing the model. We are now finished with the throat model, so you can save it and remove it from the session.

Default Layer Setup

You may sometimes need special layers created only for a single model. However, if you find you are consistently using the same layer setup, it makes sense to create a set of default layers that will be used for all your modeling (just like using a part template). There may even be company standards for layer definitions that everyone should use all the time. There are two ways to create a default set of layers (you can use either or both):

1.   Add the layer definitions, using Rule selection, to your commonly used part and assembly templates.
2.   Set up the layer definitions in your *config.pro* file.

As an example of the first method, create a new part **layertest** using one of the built-in part templates such as **mmNs_part_solid**. Open the layer tree for this part (see Figure 27). It contains a number of layers, mostly dealing with datum features. Notice that some are rule-based.

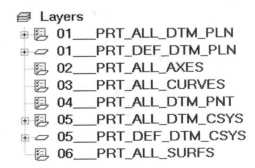

Figure 27 Default layers in **mmNs_part_solid**

To use the second method, you must edit your *config.pro* file as follows. Select

File > Options
Configuration Editor

Click *Add*, then in the **Option** data field, type in the option name *DEF_LAYER*. In the **Value** field, open the pull-down list (it is pretty long!) and select *LAYER_CURVE.* Now you must manually add the text "curves" to the value setting. This text will identify the layer in the layer tree. Repeat this procedure to create the entries shown in Figure 28. The next time you create a new part (notice the blue wand icon),

```
|===================================
|== DEFAULT LAYER DEFINITIONS
| DEF_LAYER       LAYER-TYPE     layer_name

DEF_LAYER       LAYER_CURVE curves
DEF_LAYER       LAYER_FEATURE all_features
DEF_LAYER       LAYER_NOGEOM_FEAT nogeometry_feats
DEF_LAYER       LAYER_ROUND_FEAT all_rounds
```

Figure 28 Some default layer assignments in *config.pro*

these four layers will be created automatically. Select *OK* in the **Options** window to save these new settings to your *config.pro* file.

The **nogeometry** layer is a good way of catching any and all items that are "hide-able." The **all_rounds** layer is a good way of selecting all rounds at once, perhaps for suppressing them.

To see if the default assignments are working, create a new part **layertest2** with the empty template. Open up the layer tree - it doesn't show any layers (yet!). Create the default datum planes and the layers ALL_FEATURES and NOGEOMETRY_FEATS will appear. Now create a single datum curve (say a circle in the TOP datum), then a protrusion. Round one of its edges. Verify that all of these features are appearing where they should in the layer tree (Figure 30).

A difference between these two methods is that layers in the part template are present right from the start, even if they contain no features. Layers created using the configuration file only show up if the appropriate feature is actually created.

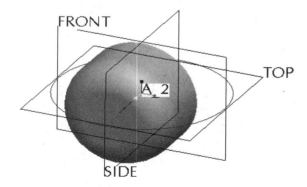

Figure 29 Test part **laytest2**

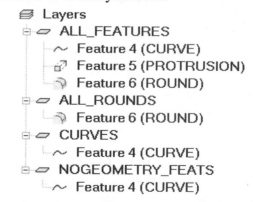

Figure 30 Default layers using *config.pro* options

As you can see, layers allow you a lot of flexibility in organizing the features in your model. The functionality we have seen here for a part also applies to assemblies. There, we can assign different components to different layers. The layer display works somewhat differently in assemblies, as it also interacts with the shading and hiddenline display of components. You might like to experiment with this on your own.

This lesson has introduced you to the operation of Pro/PROGRAM and Layers. The former allows an easy interface for users not familiar with how the model was constructed to create variations on the model geometry by specifying input values for variables and to control the regeneration sequence. Pro/PROGRAM also presents an easy way to create family tables.

The use of Layers will not only give you more control over the screen display of the model but will assist you in organizing the model so that operations on groups of related features are simplified.

In the next lesson we will look at some advanced techniques used in the creation of drawings.

Questions for Review

1.   What are the five content areas in a design file?
2.   What is the extension associated with a design file for (a) a part and (b) an assembly?
3.   What are the three types of variables that can be used as input to a design?
4.   Give examples for the syntax used for the following elements in a design:
      a.   user input of variables
      b.   a relation
      c.   a conditional branch
5.   In a boolean expression, what is the correct form for an equality test?
6.   What is meant by "incorporating" the design?
7.   What is the difference between reading a design from the model and from a file?
8.   When does a design file execute?
9.   Where can a conditional branch be placed in a design file?
10.   How does Creo react if there is a syntax error in the design file?
11.   How do you create the generic part in a family table for a part which includes a design?
12.   Why is it dangerous to use "*" in a family table for a part driven by a program?
13.   What are the two primary functions of layers?
14.   What types of entities can be organized in layers?
15.   Why can a parent feature be hidden without affecting its children?
16.   Describe four ways of selecting items to be added to a layer.
17.   What is meant by a non-geometric feature?
18.   What happens if you hide a layer containing both geometric and non-geometric features?
19.   Describe the purpose of the default layers.
20.   What two pieces of information are required to create a default layer?
21.   Identify and compare the two methods for creating default layers. In particular, describe when the default layer assignment is active.
22.   Can you delete or rename a default layer that is assigned when Creo is started up?
23.   How can you create a layer that will hide all non-geometric features at the same time?

Project Exercises

Only three parts to do in this lesson. These are pretty straightforward again - in one case trivial! We will be using Pro/PROGRAM when we get to Lesson 8. The parts are all related to the front wheel of the cart:

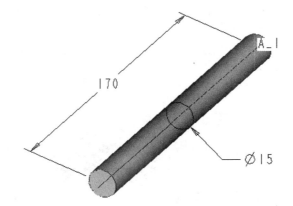

PART: *front_axle*

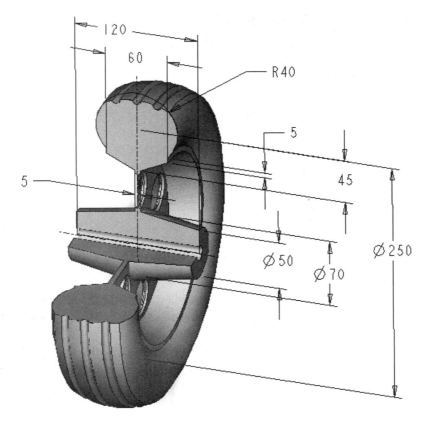

PART: *front_wheel*

The bracket below is made from 5mm thick plate (pretty hefty!). The plate is thickened quite a bit where the wheel axle will go through. The 4 bolt holes at the top are on an 80 mm diameter bolt circle centered on the top surface. You may have to estimate some dimensions here, and modify them later when we start assembling the cart.

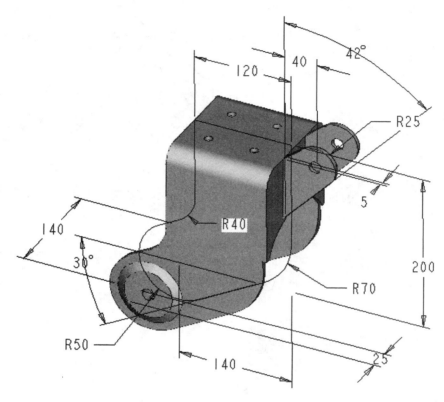

PART: *front_wheel_brack*

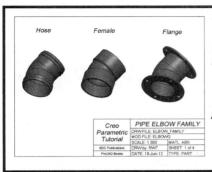

Hose   Female   Flange

Creo Parametric Tutorial	PIPE ELBOW FAMILY	
	DRW FILE: ELBOW_FAMILY	
	MOD FILE: ELBOW.G	
	SCALE: 1.000	MATL: ABS
SDC Publications	DRW by: RWT	SHEET: 1 of 4
ProCAD Books	DATE: 18-Jun-12	TYPE: PART

Lesson 7

Advanced Drawing Functions

Synopsis

The drawing setup file; dimension symbols; draft dimensions; tools for creating draft entities; drawing formats and parameters; tables and repeat regions; multi-model drawings; multi-sheet drawings; drawing templates

Overview

In this lesson we will examine a number of advanced features and techniques relating to the creation of 2D drawings[1]. We will concentrate mostly on drawings of parts, but many of the ideas will carry over to assemblies. It is assumed that you understand the basics of creating and orienting views, and showing dimensions using the tools in the drawing ribbon. We will examine the drawing setup file and see how to create a default setup. We then move on to techniques involved in detailing and interpreting the dimension symbols on the drawing. There will be a very brief introduction to the tools that you can use to create draft entities on the drawing. We will spend some time discussing tables, formats, and repeat regions, all of which can make creation of production and working drawings a much more efficient task. We'll look into multi-sheet drawings and creating drawings with more than one model. This topic is particularly relevant to assembly drawings (which we will deal with in the next lesson). Finally, we'll have a look at the creation of drawing templates.

Drawing Setup Files

There are well over a hundred settings that control the appearance and functionality of a drawing. Although some of these are contained in *config.pro*, most are included in a drawing setup file. When a drawing is first created, it contains the settings in a default setup file. Different drawings, even in the same session, can have different setup files, or

---

[1] Creo has tools for creation of annotations directly on the 3D model. This includes full GD&T capability according to standards. See the online help at
Fundamentals > Model-Based Definition
Although 3D documentation is becoming more common, the use of 2D engineering drawings remains a common industrial practice.

variations of a single file. In this part of the lesson, we will look at some of the options for determining the setup file, how to change the settings contained in it and apply it to the drawing, and how to create your own default setup file.

Start Creo and create a new drawing:

File > New > Drawing | [tut_drw_test]

Deselect the option "Use default template" and select *OK*. The **New Drawing** dialog window appears. Leave the Default Model as **none**, check the radio button **Empty**, leave orientation as **Landscape**, set the size to **A**, and select the *OK* button.

The default drawing settings for your installation could be determined in a number of ways. To see your current settings, starting in the top pull-down menu, select

File > Prepare > Drawing Properties

and then in the **Detail Options** line, select *Change*.

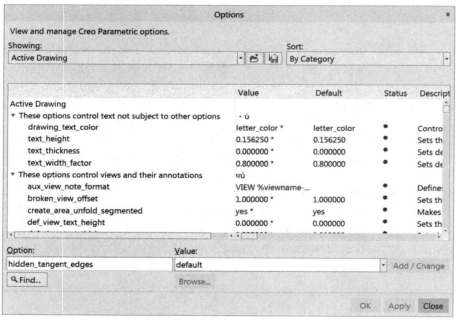

Figure 1 The drawing **Options** dialog window

This brings up the **Options** editor with your system's default drawing setup file; see Figure 1. This dialog window operates in much the same way as the window for dealing with your Creo configuration file *config.pro*. The first two columns contain the setting names and the current setting values. Unlike *config.pro*, where you (usually) only see options that differ from the default, in this drawing setup window you see all the options - there are around 130 settings listed in this file. Browse through them to see some of the options. Notice that the third column shows you a default value. You will find options for text height, text thickness, view options, arrow styles, cutting line styles, dimension appearance options, tolerance display, leaders, axes, and so on. Unless your system is set up differently, the default settings are ANSI standards.

All the options are described in the on-line help at

> *Detailed Drawings > Detailed Drawings*
> *Setting Up and Configuring Drawings > Detail Options*.

Note the **Sort** methods (see the top right corner of the window). Change this to **By Category**. This is a useful way to organize settings.

A standard Creo installation contains a number of drawing setup files that adhere to various standards. The default file *prodetail.dtl* is for ANSI standards. Standard setup files are available for ISO and DIN. To see some of the other standard setup files, select the *Open File* icon in the **Options** dialog window. The **Open** dialog window appears. Select **Drawing Setup Directory**. This will be something like the directory *<loadpoint>*/text. Select the *iso.dtl* file, then *Open*. This is a setup for a metric drawing - browse through the listed settings. Note the option ("drawing_units") is in millimeters and the projection type ("projection_type") is first angle, as is common in Europe. All options like text height, arrow length, and so on are in millimeters. In the fourth column (**Status**), it is easy to identify the ISO settings that are different from the default ANSI settings where units are inches (note the bright green stars for newly applied ISO settings *vs* dark green circles for current ANSI settings).

Once a setup file has been brought into a drawing, it will override any current settings. You will have to *Apply* the current values to the drawing to see the changes.

Retrieve the *prodetail.dtl* setup file from the **Drawing Setup Directory**. Set the sort method to *As Set*. Change the text height to **0.125** and press *Add/Change*. Then *Apply*. This setting will apply only to the current drawing. To make this or any other change "permanent" (i.e. usable by other drawings) we need to save the new setup file:

> *Save A Copy* (use the icon in the **Options** window)

Select your working directory and enter a name *def_ansi* (the extension dtl is added automatically), then *OK*. The file is now stored in the current working directory. Select *Close*, then also *Close* the **Drawing Properties** window.

Your default drawing setup file is determined by a setting in *config.pro*. If this setting is omitted, the default is *prodetail.dtl*, which produces ANSI standard (English units) drawings. To change the default setup file to the one we made above, add the following to your *config.pro*:

```
drawing_setup_file   <your working directory>/def_ansi.dtl
```

Make sure the path to the *dtl* file matches your normal startup working directory. Leave the Options window with *OK* and on the way out save the *config* file. This will ensure that Creo is always able to read this file. On a network with several users, this *dtl* file (and any others in common use) can be placed in a directory accessible by all the users. Each user's *config.pro* setting should point to this directory (if different from the default) using the configuration option `pro_dtl_setup_dir`.

To see if the new *config* setting is working, erase the current drawing and create a new empty drawing. Examine the setup file for this new drawing using *File > Prepare > Drawing Properties*. You should find it contains the modified text height that we set above.

You might like to experiment with other drawing settings as we go through this lesson, such as arrow style, text font, default view states, and so on. Erase this drawing from the session.

Detailing: Dimension Symbols and Draft Entities

In this section we will look at some of the commands and functions for detailing on the drawing. As you know, when a drawing is being created from a model, the dimensions in the model can be managed on the drawing (i.e. placed or removed) using the *Show* command. These are called, appropriately enough, "shown" or "driving" dimensions. These dimensions have *bidirectional associativity* with the model - the numerical value can be changed in either the model or the drawing and the other will update automatically - they "drive" the model[2].

One of the ramifications of using a constraint-based modeler such as Creo is that some of the geometry in the model may have been determined by constraints set up when the various features were created, particularly in Sketcher. These include alignments, equal radii, equal length, perpendicularity, and so on. Thus, there may not be any dimension values available to be shown on the drawing for some features. Although the solid modeler is happy with this (since it knows what the constraints are), the person trying to read the drawing may not be aware of these constraints and therefore would perceive that some dimensions on the drawing are missing. The main subject of this section is to look at how these additional dimensions can be created and manipulated in the drawing.

We will explore this subject by creating a simple plate model, shown in Figure 2, and then creating a drawing. Start a new part called **dim_plate** using your part template from Lesson #1 or using the built-in **inlbs_part_solid** template. The plate is a simple one-sided protrusion off the FRONT datum plane. The sketch is shown in Figure 3. Note how few dimensions are required in Sketcher due to the constraints that have been used to define the lines along the top of the plate. The plate thickness is **0.5**. That's all there is to this model. Save it.

Now, create a drawing using the *File > New* command (Ctrl-N). Select the *Drawing* radio button, enter the name of the drawing as **dim_plate**, deselect the "Use default template" box, then *OK*. Set the template to **Empty**, choose *Landscape* orientation, and specify an **A** size sheet. Select *OK*.

---

[2] Reference dimensions created in Sketcher can also be manipulated on the drawing with *Show*. These, of course, do not drive the model.

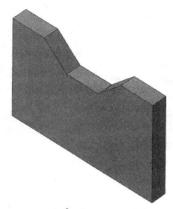

Figure 2 The plate model   **Figure 3** Sketch for plate model

Add the front view of the model to the drawing. In the **Layout** ribbon, **Model Views** group select

> *General View > OK*

or in the RMB pop-up select *General View > OK*. Click in the center of the screen and orient the view using the defined model view name *Front*. Close the **Drawing View** menu. Change the sheet scale (by double clicking on the scale value at the bottom of the screen) so that the view is a reasonable size (try **0.5**). Turn off the datum plane display.

Drawing Dimensions

To show the driving dimensions, select the **Annotate** tab in the drawing ribbon. Then either pick on the drawing view, select it in the drawing tree, or select the feature in the model tree at the bottom. Then in the RMB pop-up menu select:

> ***Show Model Annotations***

In the **Show Annotations** window, put a check beside both dimensions and *OK*. You will see just the two dimensions used in Sketcher. Use *Cleanup Dimensions* in the drawing ribbon (**Edit** group) to clean up the spacing[3]. See Figure 4. To see the symbolic names for these dimensions, select *Switch Dimensions* in the **Format** ribbon or RMB pop-up menu. The driving dimensions created in Sketcher appear (Figure 5) as *d1* and *d2*. Your symbolic names may be different.

Would you send this drawing down to the shop? Probably not. Clearly, there are insufficient dimensions on this drawing to define the notch along the top edge. We need to put some additional dimensions along the top of the part. The dimensions we are about to add to the drawing are called "created" or "driven" dimensions. Note that these are not the same as "reference" dimensions, which mean something different.

Switch the dimensions back to their numeric values with *Switch Dimensions*.

---

[3] To set the number of decimal places in the dimensions, in the **Format** group select *Decimal Places* and set the desired number.

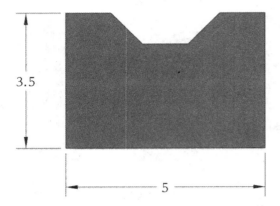

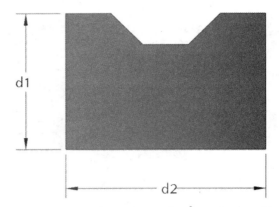

Figure 4 Dimensions created with *Show* (driving dimensions)

Figure 5 Driving dimension symbols, *d#*

Standard Dimensions using *On Entity*

Go to the **Annotate** ribbon and select (or use the RMB pop-up)

Dimension > Select an Entity

The second command is the default in the small **Select Reference** pop-up window. We use the same methods to apply dimensions as in Sketcher. Left click on the top left horizontal line. Middle click to place the dimension above the line. To dimension the depth of the notch, left click on the center horizontal line (bottom of the notch) and, with CTRL-left, on the top right horizontal line and place this dimension with a middle click. Finally, dimension the top right horizontal line and select *Cancel*. The drawing should look something like Figure 6. Now switch symbols again (Figure 7). The symbolic names appear as *ad#*, which stands for "associative dimension". This naming convention is the primary method to determine which dimensions are driving (*d#*) and which are driven (*ad#*).

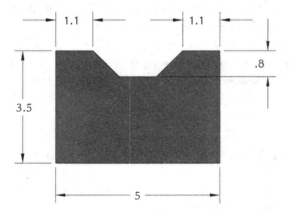

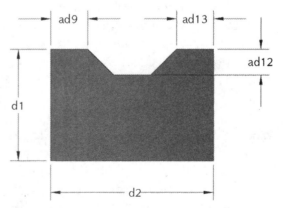

Figure 6 Draft dimensions applied to drawing

Figure 7 Symbol names for driving and driven dimensions

What happens if the part changes size? Use *Switch Dimensions* to get the numerical values back. Left click on the width dimension (5). Right click and select *Modify Nominal Value*. Enter a new value of **6**. We have to regenerate the model. Do that (with CTRL-G). The drawing should now appear as Figure 8. The associative dimensions all change as required - they are truly "driven" by the model. Change the width back to 5 before proceeding (double-click on the 6).

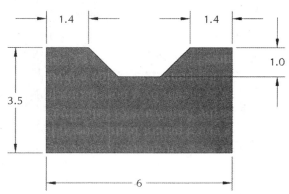

Figure 8 Model regenerated with new width

What happens if you try to change the depth of the notch? Why?

Selecting References for Dimensions

Another format for the created dimensions is the following. Erase the two created dimensions across the top by picking them (use CTRL-click) with the left mouse button (they will highlight); then use the RMB *Delete* command. Note that *Erase* just removes the dimension from the display (like *Hide*). Then select

Dimension > Select a reference

where the second command is in the **Select Reference** first pull-down list. Read the message window and click the left vertical edge of the part. This sets the reference line (in bold green) for the dimension. See Figure 9.

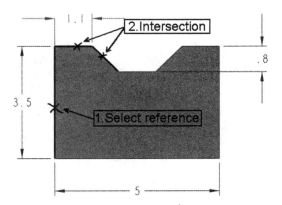

Figure 9 Creating a dimension using references

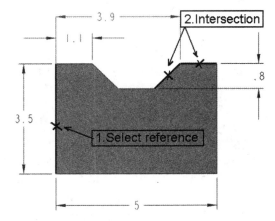

Figure 10 Creating a dimension using references #2

In the **Select Reference** window, select *Intersect* and pick the two lines (using CTRL-left) that meet at the right end of the top left edge. Middle click to place the dimension. Repeat these commands and pick the two edges that intersect on the other side of the notch. See Figure 10. Middle click to place the dimension.

The final dimensions are shown in Figure 11. If you switch dimensions, you will see that these are associative (*ad#*).

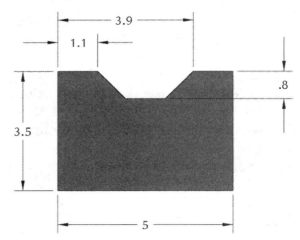

Try to *Modify* any of these associative dimensions. You can't, as stated in the message window.

Foreshadowing what is to come, try moving this view on the drawing sheet. Select the view, then in the RMB pop-up make sure that *Lock View Movement* is turned off (unchecked). You can then drag the view sideways. Left click to drop the view at the new location. All dimensions move with the view, as you might expect,

Figure 11 Associative dimensions on the top of the part using references

since the dimensions are connected to, or associated with, view geometry. Does this always happen? Before looking into that, *Lock* the view and *Save* the drawing.

Creating Draft Entities

You may occasionally want to add drawing elements (entities) to the drawing in addition to the ones displayed in the views. As an example, suppose that we want to mark an area on our plate for the placement of a small sticker. This placement location is not included in the solid model (although it possibly could be done using a sketched curve), so we have to add it here. Entities to define the placement location for the sticker can be added using some sketching tools available in the drawing ribbon under the **Sketch** tab. These sketching commands are very similar to those in Sketcher. Explore the tool buttons and flyouts for a minute.

Look ahead to Figure 13. We want to create a small rectangle on the plate. Left and right edges are lined up with corner vertices on the top edge.

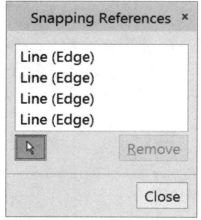

Create a couple of vertical construction lines. When you select the *Construction Line* command the cursor changes to a red cross-hair with a yellow marker at the intersection and a new window opens (it may be partly hidden). This is the **Snapping References** dialog (Figure 12). These references operate in much the same way as references do in Sketcher. Pick the pointer button under the reference collector. You may select any entity on the screen and it will become "snap-able" for new sketched entities. We want to snap to the two corner vertices, so pick on the existing drawing edges that meet at the vertices. As you pick them, they will

Figure 12 Dialog window for creating snapping references

turn orange with blue vertex highlights. (Easier to see if you are in No Hidden display mode.) As each reference is selected, it appears in the **Snapping References** window.

Middle click (once!) to get back to Sketcher. As you move the cursor cross-hair around, it will snap to the designated references.

If you middle click twice, your Snapping References are lost. If that happens, just pick a sketching tool again and reselect the references. If you accidentally pick a reference you don't want, picking on the listed reference and the Remove button becomes active.

Move the references window out of the way. First, create a couple of construction lines through the existing geometry. With **Construction Line** selected, click on one of the top vertices of the notch. The construction line will snap to vertical. Left click to create the line. Repeat for the other vertex. Note that the two construction lines are now listed as Snapping References.

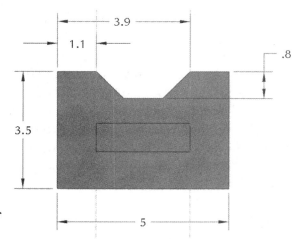

Figure 13 Draft geometry entities

Now choose the **Line** tool to create the two horizontal edges of the rectangle. As you select these pick points, notice that you may snap to constraints and alignment points (much like Intent Manager in Sketcher). These also become snap-able references. Close the two ends of the rectangle, which should now appear as in Figure 13. Middle click to leave sketching mode. The newly sketched lines appear and the references on the top edge disappear.

If the corners of the rectangle don't exactly meet, don't worry about it because we will be trimming these up in a minute or two.

Dimensions of Draft Entities

We want to create a couple of dimensions for the rectangle. In the **Annotate** ribbon, or using the RMB pop-up menu, select

> ***Dimension***
> ***Select a reference***

Pick on the bottom edge of the part and the bottom line in the rectangle. Middle click to place the dimension. Read the message window (you may have to scroll back to see something about drawing objects being automatically related to the view - we will discover what this means in a minute). Now, dimension the height of the rectangle. Both of these are associative draft dimensions (*add#*), Figure 14.

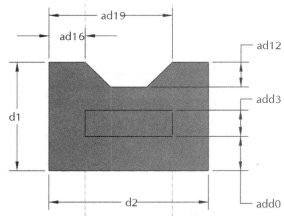

Figure 14 Dimensions created on draft entities

Check out the on-line help ("*Switching to Dimension Symbols*") for the meaning of the symbols *d#*, *ad#*, *add#*, and *dd#*.

Relating Entities to Views

To illustrate a problem with what we've just done, move the view again. Unlock the view, then click on the view and drag it sideways. *Repaint*. The rectangle and construction lines are no longer in the same place on the part, Figure 15. The view has moved, but some of the draft entities are stuck to the drawing sheet.

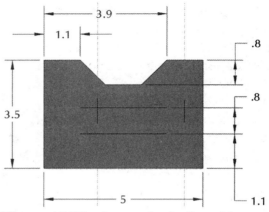

Figure 15 View moved - draft entities left behind!

We want to move the view back to its original position and have the draft entities revert to their previous position. We could use *Undo* to back up, but in order to explore some new commands, go ahead and delete these entities now (you must be in the **Sketch** group to do this). When you delete the rectangle lines the *add#*'s disappear too. Move the view back to the center of the sheet. Let's do something a bit different.

Recall that the previous rectangle was drawn one edge at a time. Create the two vertical construction lines as before (through the reference vertices on the top corners). Now middle click to accept them. In the **Settings** group select the *Chain* button. Now select *Line* tool and use the **Snapping References** dialog window to pick the two construction lines as references. Now you can create the rectangle with a series of mouse clicks (one at each corner vertex). The end vertex of each line becomes the start vertex of the next, in the style of Sketcher, and each new line rubber-bands from the start vertex. When all four edges are drawn, middle click once to end line creation, then again to accept the drawing entities. Due to the *Chain* button setting, this is exactly the way the line command works in Sketcher. Go to the **Annotate** tab and dimension the height of the rectangle as before. Your drawing should once again look like Figure 14.

We need a way to tell Creo to keep the draft entities attached to the view. This is called *relating* the entities. Go back to the **Sketch** ribbon tab, and select (with CTRL) the four edges of the rectangle (or use a selection box). The edges highlight. Then, in the RMB pop-up select

> *Relate to View*

Read the message window and click on the view, then middle click.

Now unlock the view and move it again. The rectangle lines move with the view but the two construction lines don't. It is easy to understand why - they weren't selected to be related to the view because they were not completely inside the selection box above. We don't need them any more (the rectangle is now attached to the view[4]), so delete them.

So here are the key points about related entities:

> *If a draft entity is dimensioned to a view, it automatically becomes related to it. Conversely, if a draft entity is not related to a view (either by dimension or explicit command) it will not move with the view.*

More Drafting Tools

There is quite a lot you can do with the drafting tools available in drawing mode. You can, in fact, enter this mode directly (without creating a solid model first) and create complete 2D drawings. Although Creo is not really set up for this, it does have all the necessary tools. You will normally require these tools only to "touch-up" an existing drawing. Just so that you have a look at these, some of the tools are described below. While you are doing this, scan down the command menus to see other options. You might come back later and experiment with these on your own.

Moving and Trimming Draft Entities

We want to make the rectangle a bit taller. Can we do that by changing the dimension value? Double-click on the height dimension. We cannot make this change. (Why?) Instead, we have to move the rectangle edge.

Click on the bottom edge of the rectangle. Drag it downwards. The vertical dimension values will adjust (remember that they are associative). Drag the upper line upwards. Move the lines until the rectangle height dimension is about **1.5** as shown in Figure 16, and the rectangle is about **0.8** from the bottom of the part. The corners have probably become disconnected. To restore nice corners on the rectangle, in the **Trim** group select

> *Corner*

Pick on the lower horizontal line and one of the vertical lines (using CTRL). Repeat for the other four corners of the rectangle.

While you're here, browse through the remaining tools in the **Sketching**, **Trim**, and **Edit** groups. Between these and the groups in the **Annotate** tab, you can do just about anything for creating simple 2D drawings.

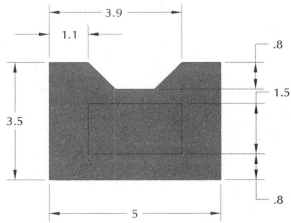

Figure 16 Final rectangle size

---

[4] However, observe what happens if you change the width of the part to 6.0 as we did previously. See if you can fix this problem.

Changing the Line Style

We don't want the rectangle to show on the drawing as a visible edge, so let's change the line style. In the **Format** group, select

Line Style > Modify Lines

and drag a box around the rectangle. Middle click. A new dialog window appears. In the Line Font pull-down list, select **CTRLFONT_S_S** and then *Apply*. If you don't like that font, try another one and *Apply* again. You can also set the color of the line. When you're happy with the line style, select *Close*. Select *Done/Return* in the **Line Styles** menu.

You can change the line style of any line on the drawing. This is handy if you bring in datum curves (perhaps as the trajectory of a sweep) and want to show them in a centerline font, for example.

Adding a Note and Hatch Pattern

Let's fill in the rectangle with a hatch pattern. Select the four sides, then in the **Edit** group select

Hatch/Fill

Type in a name like "sticker". In the MOD XHATCH menu select *Retrieve* and bring in the hatch pattern for zinc (listed under Custom Patterns). If this does not come up on your installation, just use the default hatch. Change the hatch scale and angle if you want. Then select *Done*.

Finally, let's add a note to the drawing. In the **Annotate** ribbon select:

Leader Note

Pick on the edge of the rectangle then middle click where you want the note elbow. Enter some text for the note like "PUT STICKER HERE", then middle click. See Figure 17.

Do you have to relate the hatch pattern and the note to the view? Try to move the view. What happens and why? Save the drawing and part, and erase them from the session.

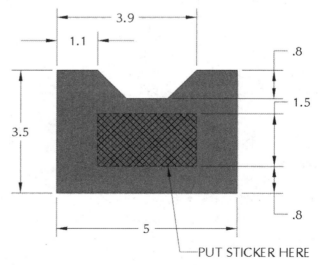

Figure 17 Final drawing of *dim_plate*

Drawing Formats and Tables

A format for a drawing contains such things as the border and title block which you use often in creating your drawings. The format can be applied either when the drawing is created or after the drawing is already underway. Formats can also be removed or replaced on drawings. Formats are defined for specific sheet sizes. You can have multiple formats defined and select the one you want for an individual drawing or sheet.

There is no default format, but there are several standard formats containing ANSI standard title blocks for several sheet sizes (*a.frm*, *b.frm*, and so on). If not installed in the default location (<loadpoint>\formats), a specific format directory can be identified in *config.pro* using a setting similar to (your path may be different)

```
pro_format_dir         c:\myfiles\proe\formats
```

You may have occasion to create your own format[5]. This section will show you the steps required to not only create the format with a title block but also to make automatic entries in the title block. This will be accomplished using a table with model and drawing parameters.

Creating a Format

In the Creo main window, select

File > New > Format | [tut_format] | OK

In the **New Format** window use the **Empty** radio button, select *Landscape*, set the size to **A** (an 8.5 X 11 sheet), then *OK*. You can close the Navigator window.

We will first place a 3/8" border around the page using some drawing tools. Select the **Sketch** ribbon and in the RMB pop-up select

Line Chain

(or use the *Line* button). The cursor will change to a red cross hair. Now right click and select (this is also in the **Controls** group in the ribbon)

Absolute Coordinates

For the first point, enter the absolute coordinates X = **0.375** and Y = **0.375** (drawing units are inches). An orange line will rubber-band out from the first vertex. For the second point, right click again and select

---

[5] Incidentally, it is a good idea to store files like this (and part templates, drawing setup files, etc) in some place other than the default installation directory under the Creo loadpoint. This is because you don't want them lost when a new version of Creo is installed.

Absolute Coordinates

and enter X = **11 - 0.375**, Y = **0.375**. Observe that calculations are allowed in the entry field.

Now to create the left vertical edge. Note that (in the Snapping References window at the upper right) the horizontal line is a snap-able reference. Start the next line by picking on its left end. For the other end, right click and select

Relative Coordinates

and enter the relative coordinates X = **0**, Y = **8.5 - 2 * 0.375**

Now we create a horizontal line across the top. Pick on the top of the left edge (it should be snap-able). Move the cursor horizontally to the right. The line should snap to horizontal and should also snap to match the length of the lower horizontal line.

For the final vertical line on the right side, you should be able to snap to existing vertices. Just pick on the right ends of the bottom and top lines. We now have a rectangular border 0.375" from the edge of the page. Middle click to leave the *Line* command.

There were a lot of mouse short-cuts here, and an inadvertent middle click may have bounced you out of drawing mode before you wanted. Just for practice, then, erase these lines and draw the border again. You can select existing lines on the sheet as references using the RMB pop-up menu and picking *Select References* at any time.

Now, on to the title block. Rather than create individual lines to produce this, we will use a table. This not only makes the construction easier but also makes it possible for the title block to fill in automatically when the format is used.

Creating a Table

The title block table will be placed by identifying a starting point and the direction in which to "grow" the table (add rows and columns). Our table will start at the bottom right corner of the border and grow upwards and to the left. Go to the **Table** tab in the ribbon. Then either in the **Table** group or RMB pop-up menu, select

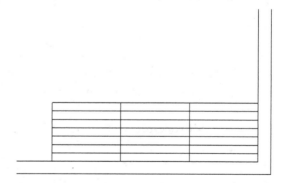

Figure 18 Creation of title block table

Create Table

In the **Direction** pane select the fourth button so the table will grow leftward and ascending. Set the **Table Size** to **3** columns and **7** rows. Set the row height at **0.25** (inches) and set the column width to **2.0**. Then click *OK*. In the **Select Point** window, pick the button on the far right (*Select Vertex*), then left click on the bottom right corner of the border we created earlier, then *OK*. See Figure 18.

Now we can modify the cells in the table by combining or merging them. Look at Figure 19 to see where we are going. First, combine the top four cells in the first column. In the **Rows & Columns** group, select

Merge Cells

Click on the top and 5th cell in the first column. This will remove the intermediate row lines. Click on the top cell in the second column, and the second cell in the third. Finally, click on the elements in the next two

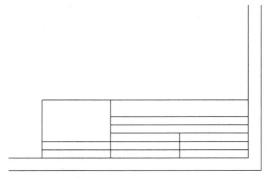

Figure 19 Completed title block table showing merged cells

rows to merge cells in columns 2 and 3, then **Repaint**. The final table should look like Figure 19. Middle click twice.

On a drawing, notes can be added to the cells in this table, a laborious and repetitive process. We can do a lot better than manual operations here, as described in the next section.

Using Drawing Parameters

We would like entries in the title block (drawing and model file names, drawing scale, sheet number, title, date, and so on) to be made automatically when the format is used in a drawing. For a preview of the final result, see Figure 21. This is done by entering drawing or part parameters in various cells in the format table. When the format is added to the drawing, the table contents will automatically display the values of the parameters. There are two types of parameters to consider: system and model parameters. In our title block, we will use both types of parameters.

System Parameters

System parameters are automatically built in to either the drawing or the model, and can be used wherever you want. System parameters you might like to use in a drawing are shown in Table II. There are others - see the online Help. Most of these parameters can be used anywhere on the drawing, for example, in a note.

We will add some of the system parameters shown in Table II to the title block. Pick in the cell on the second row from the top on the right of the table. In the **Table** group, select *Properties*. Then in the table cell enter the following text:

```
DRW FILE: &dwg_name
```

Double click in the cell below this and enter the following:

```
MOD FILE: &model_name
```

Enter the following parameters in the cells indicated in Figures 20 and 21:

```
SCALE: &scale
SHEET: &current_sheet of &total_sheets
TYPE: &type
DATE: &todays_date
```

Some of the text may overlap adjacent cells - don't worry about this. The final text will fit within the cells when the parameter value is displayed.

TABLE II System Parameters

Parameter	Definition	Parameter	Definition
&model_name	name of model	&dwg_name	name of drawing file
&scale	drawing scale	&type	model type
&format	format size	&todays_date	date format added to drawing
&linear_tol_0_0	linear tolerance values	&angular_tol_0_0	angular tolerance values
&current_sheet	sheet number of current drawing sheet	&total_sheets	total number of sheets in drawing
&pdmdb	product database of origin	&pdmrev	model revision
&pdmrl	model release level	&pdmrev:d	drawing revision

Model Parameters

Model parameters are determined by the user, either by creating the parameter in the model or in the drawing. When the format is applied to the drawing any parameters present in the format that are not defined either in the model or drawing will result in a prompt to the user to enter a value.

In the rows indicated in Figure 20, enter the following text and model parameters:

&description	
DRW FILE: &dwg_name	
MOD FILE: &model_name	
SCALE: &scale	MATL: &material
DRW by: &drawn_by	SHEET: &current_s⊦
DATE: &todays_date	TYPE: &type

Figure 20 Entering parameters in the format title block table

```
&description
DRW by: &drawn_by
MATL: &material
```

Let's make sure that the text displays nicely in the cells. Select

Text Style

and drag a selection box around the table[6]. In the **Text Style** dialog window, set the justification as follows: **Vertical** to *Middle*, and **Horizontal** to *Left*. Click on the *Apply* button and the text should all move within each cell to the requested alignment. Select *OK*. Some of cells can also have **Horizontal** set to *Center*.

We are finished creating the format. If you like, you can add additional elements to the title block as in Figure 21. This can be drawn with the tools under the Sketch menu or you can import various kinds of graphics files (IGES, DXF, TIFF and so on, even bitmaps) and place them in the format, for example a company logo. You can use either *Note* on the **Annotate** tab or the cell *Properties* command to add additional text to the title block table. The *Text Style* formatting controls will let you center the text within the cells, change fonts and text size (even color), and so on. When you are finished, save the format file in your working directory and then erase it from the session.

Using a Drawing Format

Let's try out the new format file. We'll use the *bracket* part we made in the last lesson. Open the generic of this part and make sure the units are set to millimeters (using *Same Dims*). Some parameters may already be defined in this model (*has_rib, description, modeled_by*). All our other symbolic names were dimensions. Check this out with (in the **Model Intent** group)

> *Parameters*

After confirming the existing parameter(s), assign values and/or create the required additional parameters. To create new ones, in the **Parameters** window, click on the green "+" button, then specify the parameter name, type, and contents ("value") as follows:

description	*String*	*CORNER BRACKET*
modeled_by	*String*	*[your name or initials]*
material	*String*	*STEEL*

Note that the parameter names are lowercase - these will be converted to uppercase automatically. The string values are case sensitive (what you type here is exactly what will show up on the drawing later).

Regenerate the part with the following dimension values: HEIGHT = 30, LENGTH = 20, H_DIAM = 6, HAS_RIB = Yes, and RIB_THICKNESS = 1. Save the part *bracket*.

Now create a drawing:

> *File > New > Drawing*

Call the drawing *lesson7a*. Deselect "Use default template," then *OK*. The default

---

[6] Instead of box selection, you can use CTRL-left clicks to select cells individually, then use the *Text Style* command in the RMB pop-up.

model is **bracket**. Select the radio button for **Empty with format** then (using ***Browse***) retrieve the format *tut_format* in the current working directory, then ***OK***. When prompted for the instance, open the generic part in **bracket**.

All the parameters currently in the format table are known (they are either system parameters or model parameters that we just created), with the exception of one: the format requires a value for the parameter *drawn_by*. Whenever an unassigned parameter is found, Creo will prompt you to enter a value. The title block should now be filled in as shown in Figure 21.

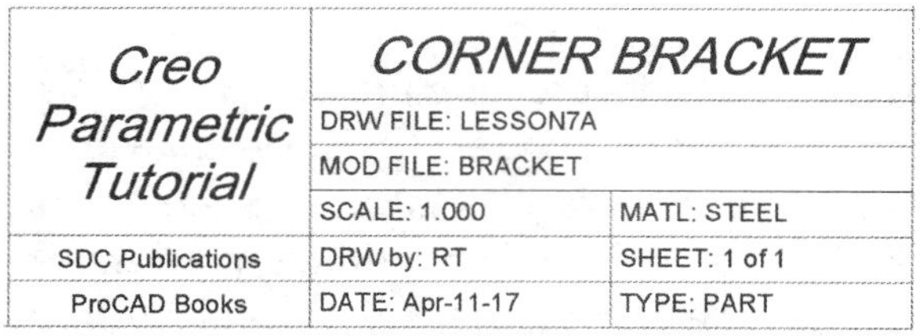

Creo
Parametric
Tutorial
CORNER BRACKET
DRW FILE: LESSON7A
MOD FILE: BRACKET
SCALE: 1.000
MATL: STEEL
SDC Publications
DRW by: RT
SHEET: 1 of 1
ProCAD Books
DATE: Apr-11-17
TYPE: PART

Figure 21 Title block on drawing with values assigned by parameters

Add a view of the bracket. In the RMB pop-up select

General View > OK

and left click to locate the view. Set up the view using the **Geometry References** option and select the FRONT datum facing right, and the TOP datum facing up. If you used our tutorial or default template, this is the LEFT view as defined in the model, in which case select ***Saved Views*** and pick the view desired. Leave the window with ***OK***. Change the sheet scale by double-clicking on the *Scale* entry at the bottom of the graphics window or in the title block. Enter a new value of **2**. Note that the title block has automatically been changed, Figure 22.

Note SCALE
Creo
Parametric
Tutorial
CORNER BRACKET
DRW FILE: LESSON7A
MOD FILE: BRACKET
SCALE: 2.000
MATL: STEEL
SDC Publications
DRW by: RT
SHEET: 1 of 1
ProCAD Books
DATE: Apr-11-17
TYPE: PART

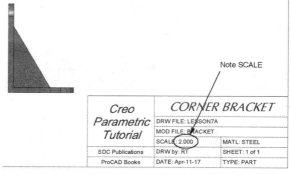

Figure 22 First view placed on drawing

To add an additional view, with the first view highlighted, use the RMB pop-up and select

Projection View

Pick to the right of the existing view. Add another projected view above the first view.

Save the drawing, but don't erase it since we need it in the next section.

Creo
Parametric
Tutorial
CORNER BRACKET
DRW FILE: LESSON7A
MOD FILE: BRACKET
SCALE: 2.000
MATL: STEEL
SDC Publications
DRW by: RT
SHEET: 1 of 1
ProCAD Books
DATE: Apr-11-17
TYPE: PART

Figure 23 All views added

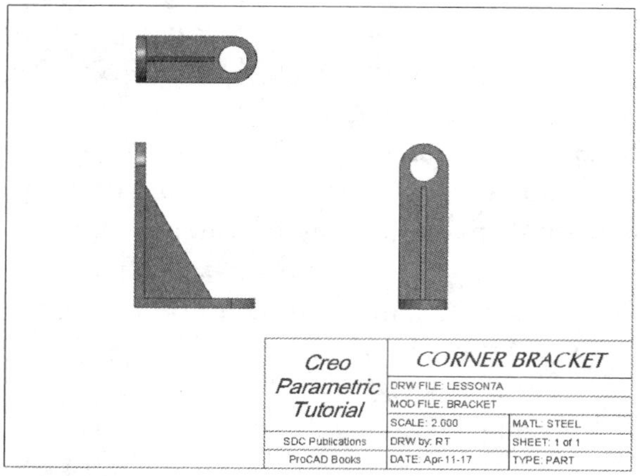

This use of a drawing format with parameters can save you a lot of time (and also leads to increased consistency) when creating drawings. In the next section, we'll construct another table that will automatically display all the members of the bracket family.

Repeat Regions

A repeat region is a portion of a table that will automatically grow and shrink according to the data in the model database. For the bracket part, we will use a repeat region to list the instances in the family table. In a later lesson, we will use a repeat region to construct a BOM (Bill of Materials) that lists (and counts) all the components in an assembly.

To construct a repeat region, we first make a table. Then we identify which cells in the table are to be repeated and what data is to be placed in each cell. As the data changes, the table will update, expand or shrink accordingly.

We will create a display of the family table instances for the bracket using two variations of the repeat region (*Simple* and *2D*). The first method is a bit more work for the family table display but gives more control over such things as individual column width and formatting within each column. The 2D repeat region requires less work but is trickier and has somewhat less control over the display.

A Simple Repeat Region

The first repeat region exercise is based on the table shown in Figure 24. To create this, go to the **Table** ribbon and using the RMB pop-up menu select

> ### *Insert Table*

Pick the second button (***Leftward, Descending***) for creation direction, then set **7** columns and **2** rows. Leave the row height at 0.25, and set the column width to **6** characters, then select ***OK***. To locate the table, we cannot use the *Vertex* option as we did for the title block table (any idea why?), so click the button for ***Absolute Coordinates***. Then enter X = **10.625**, Y = **8.125** (the top right corner of the drawing). *Merge* the cells in the first two columns of each row, as we did for the title block. The table now appears as in Figure 24.

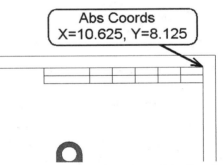

Figure 24 Creating the table for the repeat region

If you accidentally create too few columns, you can use the ***Add Column*** tool in the **Rows & Columns** group. Just follow the prompts in the message window. If the row height or column width are not quite what you wanted, you can select a cell and pick ***Height and Width*** in the RMB pop-up to change it.

Now add some text to the top row to serve as column headers. Double click the first cell in the top row (column 1). Enter the text `Part ID`. Enter the following text in the indicated columns (see Figure 25):

column 2	`Height`
column 3	`Length`
column 4	`Hole`

column 5   Rib?
column 6   Thick

The second row will be used to define six simple repeat regions. We need separate repeat regions for each cell since we are going to apply filters later, which operate over an entire repeat region. We will need separate filters in each column; hence, they must be separate repeat regions. In the assembly BOM we'll create in the next lesson, we will define an entire row as the repeat region. In the **Data** group in the **Table** ribbon select

Repeat Region > Add > Simple

As you do the following, watch the message window. Click on the first cell in the second row. This shows the starting cell of the repeat region. Click again in the same cell to show the finishing cell of the repeat region. Repeat this for all the other cells in the row (double click in each cell). These are pretty simple single-cell repeat regions. We will see later that repeat regions are not always single cells. Select *Done*.

Now we need to add some parameters to each cell. Click in the first cell on the second row; it highlights. In the RMB pop-up, select *Report Parameter*. Now pick the following sequence of menu options:

fam.. > inst.. > name

You should see the characters *fam.inst.name* in the cell. As this region eventually grows (downward), it will display the names of all the instances in the family table.

In each of the other cells on the row, use the following:

fam..> inst.. > param > value

It is probably best to work right to left here, since the text will overlap into adjacent cells.

Now we want to generate the table. In the **Data** group (or RMB pop-up) select

Update Tables

The table expands (Figure 25) to include all information in the family table. Creo does not know yet how to organize the parameter value information into the various columns, so it puts the entire table of values into each column. Recall that the display parameter for each repeat region is just set to *value* without specifying which parameter, so Creo puts them all in there! The repeat region does not know about the contents of the top row of the table.

Part ID	Height	Length	Hole	Rib?	Thick
B3020-6-R1	30	30	30	30	30
B1008-3	20	20	20	20	20
B3015-5-R2	6	6	6	6	6
	TRUE	TRUE	TRUE	TRUE	TRUE
	1	1	1	1	1
	10	10	10	10	10
	8	8	8	8	8
	3	3	3	3	3
	FALSE	FALSE	FALSE	FALSE	FALSE
	1	1	1	1	1
	30	30	30	30	30
	15	15	15	15	15
	5	5	5	5	5
	TRUE	TRUE	TRUE	TRUE	TRUE
	2	2	2	2	2

Figure 25 Table after first update and before filters

To clean up this table, we need to specify what information to keep in each column, and therefore what to ignore. This is done using *filters*.

Repeat Region Filters

To tell Creo how to restrict the data displayed in each repeat region, we use a filter defined for each region in the table (this is why we needed separate repeat regions for each column). In the **Data** group select

> ***Repeat Region > Filters***

Pick on the second column (second row or below), then

> ***By Rule > Add***

Enter the filter (note the double-"=" sign for the equality test)

```
&fam.inst.param.name == height
```

and press the Enter key twice. This means the region will only display entries associated with the height parameter. Select ***Done > Done/Return*** and the column should shorten to contain data in only the first three rows (or more if you have been experimenting with the family table entries). Repeat this procedure for each of the remaining columns using the following filters:

column 3	`&fam.inst.param.name == length`
column 4	`&fam.inst.param.name == h_diam`
column 5	`&fam.inst.param.name == has_rib`
column 6	`&fam.inst.param.name == rib_thickness`

When the final filter is entered, the table should shrink to just enough rows to show all the current instances in the family. Select ***Done*** in the **TBL Regions** menu.

Draw a selection box around the table, then go to the **Format** group and select

> ***Text Style***

Part ID	Height	Length	Hole	Rib?	Thick
B3020-6-R1	30	20	6	YES	1
B1008-3	10	8	3	NO	1
B3015-5-R2	30	15	5	YES	2

Figure 26 Completed repeat region

Set horizontal justification to **Center**, and vertical justification to **Middle**. Select an alternate font if desired. Click the ***Apply*** button, and select ***OK***. The text should appear as shown in Figure 26. Notice that the **has_rib** parameter is showing as **Yes/No**. If you prefer to see **TRUE/FALSE**, set the drawing setup file option (this is actually the default)

```
yes_no_parameter_display    true_false
```

This seemed like a lot of work. However, like a lot of Creo tasks, the work is in the planning and setup. The beauty of this table is that we can now go back to the model and create new instances (or delete old ones), and the table in the drawing will always be automatically brought up to date.

Incidentally, if you open the **Drawing Tree** in the **Navigator**, you will see both tables listed under *Sheet 1* (the **Table** tab in the ribbon must be selected). Recall that the first table is the title block in the format.

Now that we have seen a Simple repeat region, we're going to do this all over again using a 2D repeat region. Draw a selection box around the entire table, or pick the second table in the **Drawing Tree**, then in the RMB pop-up select *Delete*.

A 2D Repeat Region

This form of repeat region is ideal for displaying family tables. It requires a bit more thought and planning to set up but requires fewer commands and keystrokes. There is somewhat less control over formatting within the table.

The 2D repeat region for the family is based on a 2 X 2 table. The region will automatically expand in rows and columns depending on the data in the family table. The 2D repeat region is specified using an outer region (the 2 X 2 table) and an inner region (in this case a single table cell). The definition of the repeat region also depends on how the original table is set up, that is, in which directions it expands (Ascending or Descending, Leftward or Rightward). In the following, we will set up the table to expand downwards and to the right, as illustrated in Figure 27. Other variations of this are possible. The thing to remember is that not all 2 X 2 tables are created equal, and the definition of the 2D repeat region must be consistent with the expansion directions specified for the table. Another way we could set up the table is shown in Figure 28.

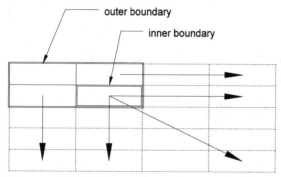

Figure 27 A 2D repeat region expanding down and to the right (Descending Rightward table)

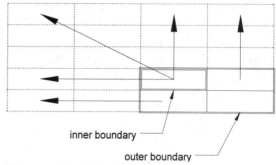

Figure 28 A 2D repeat region expanding up and to the left (Ascending Leftward table)

With the **Table** ribbon active, use the RMB pop-up to select

Insert Table

Select the first button (***Rightward, Descending***), set **2** columns and **2** rows, row height 0.25, column width **1.25**, then *OK*. To locate the table, pick ***Absolute Coordinates*** and enter X = **5.625**, Y = **8.125** for the top left corner. This location is so that the table will just reach the right border when it expands to its full width. Adjust the width of the second column to **0.75** with the command in the **Rows & Columns** group. We now have the 2 X 2 table as shown in Figure 29.

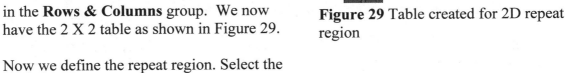

Figure 29 Table created for 2D repeat region

Now we define the repeat region. Select the button in the **Data** group

> ***Repeat Region > Add > Two-D***

For the cells defining the outer boundary, click the top left and lower right cells (see Figure 27). For the inner border, click the lower right cell. All lines will highlight and the message window will indicate that the repeat region has been defined. Select ***Done*** in the **TBL REGIONS** menu.

We only have to enter three parameters into the table. These are shown below. Compare these entries to the expansion directions of the repeat region shown in Figure 27.

	fam.inst.param.name
fam.inst.name	fam.inst.param.value

Pick the lower left cell and use the RMB ***Report Parameter*** command to define the following text

> ***fam.. > inst.. > name***

In the top right cell, enter the following

> ***fam.. > inst.. > param..> name***

and in the lower right cell enter

> ***fam.. > inst.. > param.. > value***

Now select ***Update Tables*** in the **Data** group (or use the RMB pop-up). The table expands over to the right border and fills in with the data. This is really easy - no additional filters or other adjustments are necessary to create the table. The only problem is that the column heading for the last column is a bit too long to fit within the column, and the actual column headings are identical to the model parameters (which means we should not create model parameters with names that would be hard to interpret).

Draw a selection box around the table. In the RMB pop-up menu, select *Text Style*. Use the *Select Text* button to copy the text style from the title block. Set the horizontal and vertical justification to **Center** and **Middle**. Select *Apply > OK*.

To change the order of the columns to match the order in the family table, select

Repeat Region > Sort Regions

and click on the table. Then select the radio button *No Default > Done*. The columns will rearrange to match their order in the family table (rather than alphabetical).

Let's create a new instance of the part. Recall that this part is driven by Pro/PROGRAM. Go back to the window with the generic part *bracket*, and select

Regenerate > Enter

Check the height, length and rib thickness variables. Enter new values of **25**, **25**, and **2**, respectively for these. The part regenerates with these new values. To put this new part into the family table go to the **Model Intent** group and select

Program > Instantiate

and enter a new instance name **B2525-6-R2**. If you cannot use *Instantiate* in the **Program** menu, just open the family table with *Tools > Family Tables* and enter the instance manually.

Create a final instance, **B1515-5** (smaller hole size and without the rib) and instantiate it. Go to

Family Table

to see the new instances in the family table. To change the order of presentation in the **has_rib** column, select the **has_rib** column, then *Tools > Sort > Ascending* (or *Descending*) and *OK*.

Now switch to the drawing window. The table has automatically updated to show the new instances. Return to the drawing and select the *Update Tables* in the RMB pop-up. See Figure 30.

	height	length	h_diam	HAS_RIB	b_thickness
B1008-3	10	8	3	NO	1
B1515-5	15	15	5	NO	2
B2525-6-R2	25	25	6	YES	2
B3015-5-R2	30	15	5	YES	2
B3020-6-R1	30	20	6	YES	1

Figure 30 Family table on drawing updates automatically for new instances

Note that the drawing views are showing the latest version of the part. Back in the part window, restore the part to the generic dimensions shown in the bottom line of the table (for **B3020-6-R1**). Save the part (required to save the new family table entries) and drawing.

Displaying Symbolic Dimensions

The last thing we want to do for the bracket drawing is to identify the parameters (height, length, and so on) on the drawing views. We want to show the name indicated in the table, and not the actual dimension value. In the Drawing window, go to the **Annotate** ribbon and select *Show Model Annotations*.

CTRL-click on the circular holes in the right view and top view, then on the rib in the right view. Look ahead to Figure 31 and in the **Show Model Annotations** window check the four dimensions with the symbolic names shown on the drawing. These are for the length, height, hole diameter, and rib thickness dimensions; they may not show immediately on the desired view. Accept them with *OK*. They will appear as numeric values.

Click on the dimension value for the hole height. If necessary, use the RMB pop-up menu to select *Move Item to View*, and place the dimension in the right view as in Figure 31. Select *Dimension Text*. The numerical value of the dimension is displayed because of the @D specification in the dimension text. Change this to @S and accept the dialog. This changes the dimension to show the parameter name as a string. Do the same for the hole diameter, the length dimension, and the rib thickness dimension. Change the layout and text style of each dimension as desired. The final drawing should look something like Figure 31.

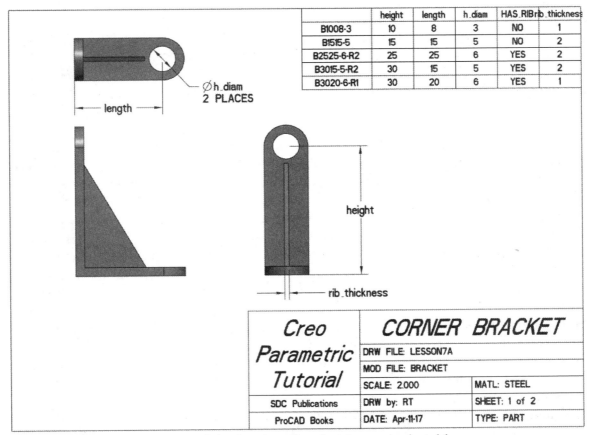

	height	length	h_diam	HAS_RIB	rib_thickness
B1008-3	10	8	3	NO	1
B1515-5	15	15	5	NO	2
B2525-6-R2	25	25	6	YES	2
B3015-5-R2	30	15	5	YES	2
B3020-6-R1	30	20	6	YES	1

Creo Parametric Tutorial	**CORNER BRACKET**	
	DRW FILE: LESSON7A	
	MOD FILE: BRACKET	
	SCALE: 2.000	MATL: STEEL
SDC Publications	DRW by: RT	SHEET: 1 of 2
ProCAD Books	DATE: Apr-11-17	TYPE: PART

Figure 31 Finished drawing of the bracket showing instances in table

We are finished with this brief introduction to the use of repeat regions. There is lots more you can do with them (including nesting of repeat regions), and you may want to experiment with these capabilities. As mentioned above, we will utilize a repeat region a bit later to produce a BOM for an assembly.

Save the drawing and erase it from the session. If you haven't already, also save the bracket part and remove it from the session. Now is a good time to take a break!

Multi-Model Drawings

It is very easy to set up a drawing to display more than one model on a single sheet. This is particularly useful for drawings of related components, as in an assembly or family table. This will be demonstrated by creating a drawing showing three of the instances in the family of elbow parts we made in an earlier lesson. Note that it is not necessary for the models to be related (as in instances of a family table) but may be independent parts. If you have not completed the elbow part, you can substitute other parts as desired, preferably one with a family table.

Open up the part **elbowg.prt**, being careful to bring in the generic of this part. We must make a few modifications to the model to go with the steps that follow. Make sure the model contains the following two parameters:

```
material      String      ABS
description   String      PIPE ELBOW FAMILY
```

As you recall, these parameters are used in the drawing format we created earlier in this lesson.

Now add a new end type ("female") using a revolved protrusion and a revolved cut. See Figures 33 and 35 for a rough idea of the geometry. Add these features to the bottom of the model tree, *being very careful about your feature creation and sketching references.* (Neither the protrusion nor the cut should reference the HOSE or FLANGE features.) The safest way to set this up is to use only the part datum planes as references, with a couple of additional datums at the opposite end of the part to fill the role of RIGHT and FRONT for features copied to that end. Also, recall that the shell thickness dimension symbol is "thick". We want the diameter of the cut to be the same as the elbow (diameter symbol "diam"); the outer diameter of the revolved protrusion should be set using the relation "diam + 2*thick". Make sure that you set up a parent/child relation between the protrusion and the cut. Eventually, we want all the related geometry to suppress when the first revolved protrusion is suppressed in the family table. Name the revolved protrusion "FEMALE." Copy the protrusion and cut to the other end of the elbow (make sure these are dependent copies). Add the appropriate entries to the family table for the part and edit the family table to define the three instances shown in Figure 32. Note that the names of the instances have also changed. IMPORTANT: *Verify* the table and *Preview* all instances to make sure you are getting what you think.

Type	Instance Name	Common Name	d2 ANGLE	d6 DIAM	d7 THICK	F128 HOSE	F281 FLANGE	F851 FEMALE
	ELBOWG	elbowg.prt	30.0	20.0	2.5	Y	Y	Y
	E-30-40-3-G	elbowg.prt_INST	30.0	40.0	3.0	N	Y	N
	E-30-40-1-H	elbowg.prt_INST	30.0	40.0	1.0	Y	N	N
	E-30-40-2-F	elbowg.prt_INST	30.0	40.0	2.0	N	N	Y

Figure 32 Family table for generic part *elbowg*

We are going to place section views on the drawing. The sections can be created in drawing mode, but it is more convenient to do this in part mode. In the **View** ribbon, *Section* pull-down menu, select

> *Y Direction*

If it doesn't exist already, a default coordinate system (CS0) will be created. The section cutting plane will be normal to the Y axis and aligned by default with the TOP datum. Go to the **Properties** tab in the dashboard and change the section name to "*A*". We can modify the hatching (spacing and angle), but we'll leave that for the drawing. Close the **Section** dashboard and note the section entry added at the bottom of the model tree. Select the section entry and in the mini toolbar click *Deactivate*. Note the **Show Section** option in the RMB pop-up. Save the part.

Creating the Drawing

Create a new drawing called **elbow_family**. Turn off the "Use default template" option. Select the default model as **elbowg.prt** and use "Empty with format". Retrieve the format *tut_format.frm* from the working directory and select *OK*. Select the generic part. Recall that the string parameter *drawn_by* is required by the format. Enter your name or initials for this.

Observe on the bottom of the window that the type of model is *PART* and the name of the current model is *ELBOWG*. You can close the Navigator pane for now.

We are going to bring in the three instances in the *elbowg* family table and place them all on this sheet.

Setting the Active Model

The drawing has opened with the tab **Layout** selected by default. In the **Model Views** group (or use the RMB pop-up in the graphics window), select

> *Drawing Models > Add Model*

Double click to select *elbowg.prt*, then select the instance **E-30-40-1-H** and *Open* > *Done/Return*. Note the name change on the bottom of the graphics window. The model file identified on the format title block is still *elbowg* (determined when the format was placed with *elbowg* as the active part). Create a view of this active model using the RMB pop-up:

General View > OK

Pick a point on the left side of the drawing sheet. Accept the default orientation for the view. In the **View Display** category, set *Display(Shading With Edges)*, then *OK*.

To change the active model to the second instance, make sure the first view is not highlighted (no border), then using the RMB pop-up menu again select:

Drawing Models > Add Model

Select the generic *elbowg* again, and this time select **E-30-40-2-F**. This name now appears as the active model on the bottom of the window. Use the RMB pop-up to add a general view of this in the center of the sheet in the default orientation. Once again, set **Shading With Edges**.

Finally, add the model and create a general view of the third instance **E-30-40-3-G** and place it on the right of the sheet in default orientation.

Rearrange the views for spacing and add some notes as shown in Figure 33. You might also change the line style of the datum curve used to define the trajectory for the sweep used in creating the elbow.

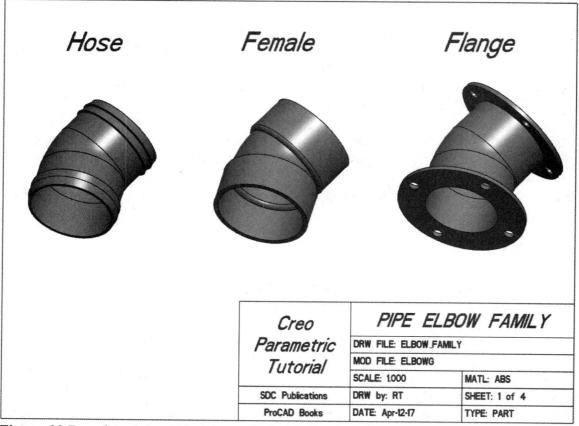

Hose	*Female*	*Flange*

Creo Parametric Tutorial	*PIPE ELBOW FAMILY*	
	DRW FILE: ELBOW_FAMILY	
	MOD FILE: ELBOWG	
	SCALE: 1.000	MATL: ABS
SDC Publications	DRW by: RT	SHEET: 1 of 4
ProCAD Books	DATE: Apr-12-17	TYPE: PART

Figure 33 Drawing sheet with three models of elbow family

A drawing can have any number of models. To see which models are currently loaded, go to the **Layout** ribbon and select

> ***Drawing Models > Set Model***

and all the currently loaded models are listed. To change the currently active model to any of these, either click on the name in the list, or just click on the view of the desired model on the sheet. Do that now, selecting the view of the hose end instance on the left. Note the name entry at the bottom of the drawing window, which should indicate that **E-30-40-1-H** is the active model. Set the active model back to **elbowg** and save the drawing.

Multi-Sheet Drawings

A drawing file may consist of more than one physical drawing sheet. As your parts and assemblies get more complicated, this will be necessary to show all the views you want to generate without requiring a large number of separate drawing files. It is very easy to navigate between sheets, switch views, tables, or draft entities from sheet to sheet, reorder the sheets, and print the sheets individually or all at once. Different sheets can have different formats (even size), different active models, and so on.

We will create three more sheets for the elbow family drawing. Each sheet will show a general view and a cross section view of one of the instances. It is permissible to have multiple models shown in each sheet of a multi-sheet drawing (just as we did above).

For the following, you need to make a change/addition to your *config.pro*. Since we are going to use the same format on multiple sheets and we want the format parameters to carry over between sheets, we must instruct Creo to add the format parameters to the model. This is done using the *config.pro* setting:

```
make_parameters_from_fmt_tables   YES
```

Adding a Drawing Sheet

To add a drawing sheet to the drawing, in the **Layout** ribbon select

> ***New Sheet***

Observe the title block and the bottom line in the graphics window. The model name indicated in the title block is still *elbowg*. Other parameters have been carried over from the previous sheet due to the *config* setting made above. Set the active model to **E-30-40-1-H**. Add a general view of this model to the right side of the sheet, accepting the default view. Now we will add the section view on the left side of the sheet using the cross-section "A" defined and named in the part. Use the ***General View*** command in the RMB pop-up menu.

Pick a center point for the drawing view on the left half of the sheet. Orient the view so that the TOP datum plane is facing FRONT, and the SIDE datum plane is facing RIGHT. You may have a saved view list, in which case, the view is the TOP view. Accept this orientation.

In the **Sections** category, select the radio button for *2D Cross-section*, then click the green "+". The section A we created before is listed. Select it; we don't have to worry about arrow display. *Apply* the section view, then turn off hidden lines in the **View**

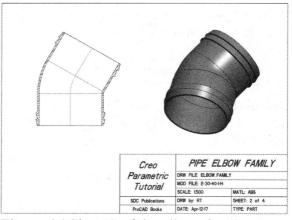

Figure 34 Sheet 2 of the elbow family

Display category. You might erase the section title and adjust the hatch spacing and angle. Change the sheet scale to **1.5**. The screen should look something like Figure 34. There are a couple of minor differences - we will fix those later.

We now add another sheet showing the next instance. This time, we will first select the new active model, then create the sheet:

> *Drawing Models*
> *Set Model > E-30-40-2-F >*
> *Done/Return*
> *New Sheet*

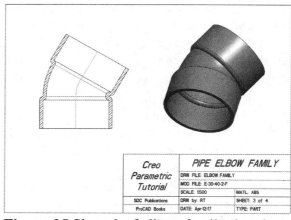

Figure 35 Sheet 3 of elbow family drawing

We are now on sheet 3 with a new active model. Note the model name in the title block is correct. Create the same two views (general and section) as we did for the previous model. Change the scale as before. See Figure 35.

We are going to add one more sheet for the flanged elbow. To remind you how the format works, we are once again going to add the sheet before we change the active model. In the **Layout** ribbon, select:

> *New Sheet*

We are on the fourth sheet with the model name **E-30-40-2-F**, that is, the model name from the previous sheet. Now change the active model using the RMB pop-up:

> *Drawing Models > Set Model*

Figure 36 Sheet 4 of elbow family drawing

and select **E-30-40-3-G** and *Done/Return*. Observe the bottom line in the window to confirm this is the active model. The entry in the title box is still for the previous instance. This is because the value of the parameter name *&model* was already set when we added the format to this sheet. We will come back in a minute to fix this.

Meanwhile, add the same two views (general and section) for the flanged elbow. Make your cosmetic changes as desired.

To update the title block, we need to replace the format (with a new copy of itself!). In the **Document** group select:

> *Sheet Setup*

or pick *Sheet Setup* in the RMB pop-up. For sheet 4, click on the format name then open the pull-down list. System formats are listed there, along with the current one. Click on TUT_FORMAT then *OK*.

Since there might be a table in the existing format that you want to keep (like a repeat region or BOM), you must confirm the deletion of the table. This prompt appears for each table (highlighted in red). In this case, we want to replace the title block table so select *Remove*. To remove all tables (and bypass the prompts) you could select *Remove All*. The format comes in with the now-correct entry in the title block, Figure 36.

Switching between the sheets is a simple matter of selecting the desired sheet tab in the toolbar below the graphics window. Go back to sheet 2 and replace the format with *Sheet Setup* so that the title block displays correctly.

You can rename the sheet tabs using the RMB pop-up menu. You can also reorder sheets and remove them. In the **Document** group, the command

> *Move or Copy Sheets*

allows you to re-order the sheets. If a view on any sheet is selected, the command

> *Move to Sheet*

in the **Edit** group lets you move views, tables, draft items and so on onto another sheet.

As you have observed, there are lots of short cuts launched using the RMB pop-up menu. You should periodically execute a RMB pop-up to familiarize yourself with the available commands. This can save a ton of time.

Finally, note that when you go to print out the drawing, you can direct Creo to produce a hard copy of all (or a subset of) sheets associated with the drawing.

Don't forget to save the drawing (set *elbowg* as the active part first)! Then erase it and all associated objects.

Creating a Drawing Template

This is the last topic in this lesson. As you probably know, a standard Creo installation includes a number of drawing templates. These are similar to drawing formats but with some important additional capability. Templates can contain the following information:

- Basic information required on the drawing that is not included with the model, such as tolerance notes, special symbols, and so on.
- Information for laying out and configuring views: view type (section views), view display (hidden, no hidden, etc.), dimensions and snap lines, balloons, and so on.
- Parametric notes that are driven by model parameters and dimensions.

Like part templates (but unlike formats), a drawing template must be chosen when a new drawing is created. In this section, we will create a new drawing template and then make a simple part to try it out. Also like part templates, a drawing template is used to control the initial creation of a new drawing. Once the drawing is created, all aspects of it can be modified however you want - you are not permanently stuck with the template layout. You can, for example, delete an unnecessary view that is defined in the template, or add a new view if necessary. A normal Creo installation includes a number of standard templates that will show, for example, the standard Top-Front-Right views.

We start the creation of a new template by creating an empty drawing using

File > New

Select the **Drawing** button, enter a name **tut_template**, deselect the "Use default template" button, and *OK*. Leave the default model area blank (**none**), select **Empty**, **Landscape**, and size **A**. A new blank drawing is created.

IMPORTANT: This is the critical step for creating a template: in the **Tools** ribbon, **Applications** group, select

Template

We can create a border and title block exactly the same way we did previously for the format exercise. See if you can do that without referring back to those instructions (remember that "practice makes perfect!"). Create the table and enter the parameters as we did

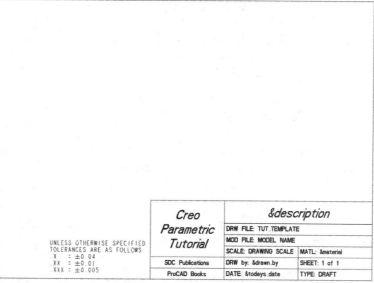

Figure 37 Template under construction - border and title complete, with tolerance note

before[7]. Adjust the text alignment, font, spacing and so on. Add a note that specifies the default tolerances for the drawing. When you are finished, the template should look something like Figure 37. Observe the parameter values waiting to be filled in. Note that if we had applied the format prior to launching the template application, we would have a different result. Can you figure out why?

Now we will add the information for laying out and configuring the views to appear on the drawing. Look ahead to Figure 41 for the desired final drawing appearance. In the **Layout** ribbon select

Template View (or in the RMB menu, ***Template View***)

The **Template View Instructions** window opens (Figure 38), the contents of which will change as we proceed with the options described below. The view name will default to VIEW_TEMPLATE_1. The **View Type** will be *General*, using the view orientation **FRONT** (as stored in the model). In the **View Options** and **View Values** areas, select the following options and settings:

<table>
<tr><td>***Model Display***</td><td>***Hidden Line***</td></tr>
<tr><td>***Tan Edge Display***</td><td>***None***</td></tr>
</table>

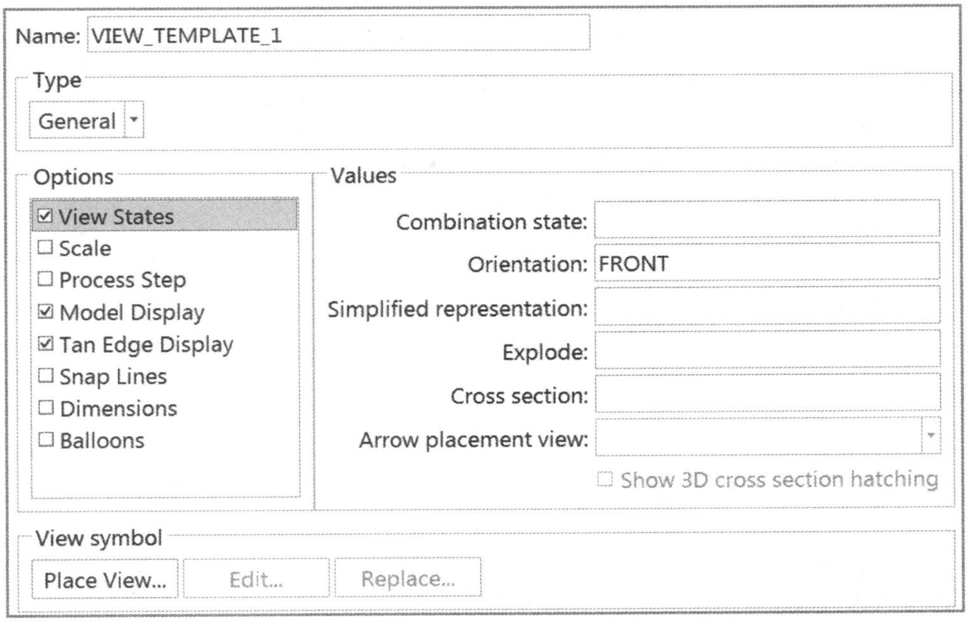

Figure 38 The **Template View Instructions** dialog window

You may want to move the dialog window over to the side a bit. Select the ***Place View*** button at the bottom and read the message window. Drag the mouse to create a view bounding box on the left-center of the drawing sheet (see Figure 39), or just left-click to designate the placement point. This will be the location of our primary (FRONT) view.

---

[7] If you are in a real hurry, then use the RMB pop-up to select ***Sheet Setup*** and select the format **tut_format** we made earlier.

Let's add another view. At the bottom of the **Template View Instructions** window, select

> *Repeat*

The View Name will be **VIEW_TEMPLATE_2**. This will be the right view, which we want to show as a dimensioned, full section. (Note that section views in templates can only be full cross sections). The **View Type** we want is *Projection* (in the pull-down list). The view will be a projection from the previous **VIEW_TEMPLATE_1**. In **View Options**, select

View States	*Cross Section: A* (we'll set this up in the part)
	Arrow Placement: VIEW_TEMPLATE_1
Model Display	*No Hidden Line*
Tan Edge Display	*None*
Snap Lines	*Number: 3*
	Increment: 0.375
	Initial: 0.5
Dimensions	*Create Snap Lines*
	Increment: 0.375
	Initial: 0.5

Now select *Place View*, and put the view symbol to the right of VIEW_TEMPLATE_1. We have asked for a projection from the front view. We do not have to be precise about the placement of the symbol, since if we are "close enough" Creo will know what to do. If we place the symbol too far away (vertically) from a true horizontal projection, Creo will complain via an error window when the template is used (and create a file **template.err** in your working directory).

Now add a top view to the drawing. Select

> *Repeat*

The view name will be **VIEW_TEMPLATE_3**. The **View Type** we want is *Projection* (in the pull-down list). The view will also be a projection from **VIEW_TEMPLATE_1**. In **View Options**, select

Model Display	*Hidden Line*
Tan Edge Display	*None*
Snap Lines	*Number: 2*
	Increment: 0.375
	Initial: 0.5
Dimensions	*Create Snap Lines*
	Increment: 0.375
	Initial: 0.5

A new button, *Set Display Priorities*, appears. This lets us choose the priority list for where Creo should place the dimensions on the drawing (i.e. which views). Leave VIEW_TEMPLATE_2 (the section view) at the top of the list. Now select *Place View*, and place the view symbol at the top of the sheet above our first view. Select **OK** and return to the main drawing window. Your drawing should look like Figure 39. Save the template in your working directory. It is stored as a drawing (*drw*) file. You can then close the window and remove the template from the session.

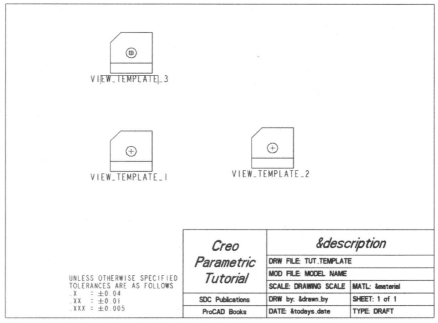

Figure 39 Completed drawing template **tut_template.drw**

We need a simple part to try out this new template. Create the part **small_tank** as shown at the right. Use the **inlbs_part_solid** part template. The part is basically a revolved protrusion that has been shelled out (thickness **0.25**). Then a flange is added using another revolved protrusion. Because we want these dimensions to appear in the right view, make the sketches for these solid features on the SIDE (RIGHT) datum. Finally, create a planar cross section (*X Direction*, named "A") on the SIDE (RIGHT) datum. Also, make sure we have parameters *description*, *material*, and *modeled_by* defined and specified in the part. Save the part.

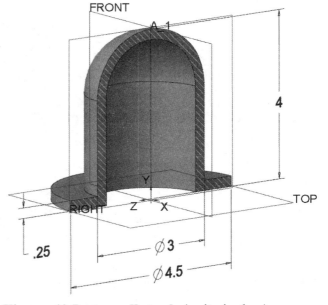

Figure 40 Part **small_tank** (units inches)

Now for the fun part! Create a new drawing called **small_tank**. Deselect the "Use default template" option, then **OK**. The default model should be **small_tank**. Select the "Use template" option, and use the **Browse** button to find the **tut_template.drw** file in your working directory. Then select **OK**. Your drawing appears as specified (see Figure 41): three views, including a section view and arrows, view display as desired, dimensions and snap lines, and a completed title block. Pretty simple! All you have to do is provide some information for the **drawn_by** parameter. You will probably still want to do some cosmetic clean-up on this drawing. Are all the usual tools available? For example, you may want to modify the sheet scale, move some dimensions to a different view, change the placement of some dimensions, move views, add center lines and axes, possibly add required associative dimensions, and so on. However, we have basically automated most of the production of this drawing.

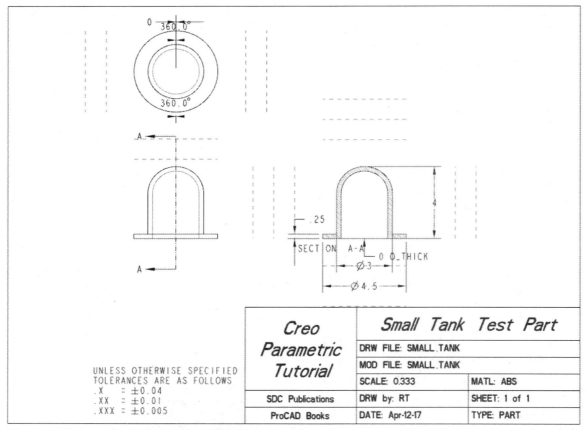

Figure 41 Drawing of **small_tank** as created by **tut_template**

For further information on drawing templates, consult the on-line help by going to the **Contents** area and selecting

> *Detailed Drawings > Detailed Drawings*
> *Setting Up and Configuring Drawings > Templates for Drawing Layout*

Summary

In this lesson, several topics relating to creation of drawings have been introduced. You should become familiar with setting and modifying the options in the drawing setup file. Draft entities may be required to touch-up or place additional information on a drawing. In Creo, you have all the tools of a 2D drawing package to do this. Don't forget that you may have to explicitly associate some entities with a view. Tables were introduced as a way to organize information on the drawing. This was applied to the creation of a format with title block and a repeat region to display family instances. A quick look at multi-model and multi-sheet drawings has introduced you to some useful methods and functions to create more comprehensive drawing sets. Finally, we looked at the creation of drawing templates.

In the next lesson we will look at functions relating to assemblies, including some more advanced drawing functions. We will finally start putting the cart together! There are three more parts to be made at the end of this lesson, though.

Questions for Review

1. Where is the default drawing setup file located on your system?
2. Where are the additional setup files stored on your system?
3. How do you identify drawing setup files (that is, what is the file extension)?
4. How do you point your system to a specific drawing setup file to be the default?
5. What commands do you issue to edit a setup file?
6. Can you create a metric drawing on a sheet whose format was created using English units?
7. What is the difference between first angle and third angle projection?
8. How do you apply a new setup file to a previously created drawing?
9. In regards to dimension types, explain the terms (and the relations between them): "shown", "driven", "driving", and "created".
10. How can some features appear on a drawing with no dimensions even if you select ***Part > Show All***?
11. What is a "reference dimension" and how do you create it?
12. What determines whether created dimensions are stored with the model or with the drawing? What is the difference in operation?
13. How do you find out the symbolic names for dimensions? How can you display these on a drawing?
14. The four kinds of dimensions symbols are *d#*, *ad#*, *add#*, and *dd#*. What do these various forms mean?
15. What is in the group of controls at the bottom of the drawing window in a multi-sheet drawing?
16. What happens if you delete a *d#* dimension?
17. Where are the drawing commands for creating draft entities?
18. Where are the commands for editing draft entities?

19. Suppose you have just created a drafted construction circle on a view of the part. Can you dimension the location of the circle relative to an edge of the part? Why?

20. Is it possible to delete a line on the drawing that is a part edge in a view?

21. Can you change the line style of a line representing a physical edge of the part?

22. When can you specify the format for a drawing?

23. How many formats can be included in a multi-sheet drawing?

24. Where are format files stored on your system?

25. Where are the standard formats containing the ANSI title blocks stored?

26. When specifying absolute coordinates of a point on the sheet is the origin at the corner of the sheet, or the corner of the border? Can you change it?

27. What options are available to set the width and height of the cells in a table?

28. How do you set the directions for growing a table?

29. How can you add columns or rows to a table after it has been created? What determines their width/height?

30. How can you remove columns or rows from a table after it has been created?

31. Can you remove a table row from a repeat region driven by a family table?

32. Suppose you want to put some text into a table. What is the difference between using the RMB *Properties* command on a cell and *Note*?

33. Explain the difference between a system parameter and a model parameter.

34. What happens if a format containing model parameters is applied to the drawing of a part that does not have those parameters defined?

35. How do you assign the parameters to a cell in a repeat region?

36. What happens in a repeat region for a family table if a parameter value in the table is "*"?

37. How can you sort a repeat region? What options are available?

38. What happens if a repeat region grows off the edge of the drawing sheet?

39. How do you modify a dimension so that it shows the symbolic name?

40. What does the dimension text "@O" do? This is a capital O not a 0 (zero).

41. Is it possible to have a model added to a drawing without showing it in any views?

42. How do you determine which model in a drawing is currently active? How can you change the active model?

43. How many models can be brought into a drawing?

44. If you remove a sheet that contains the only view of a model in a drawing, is the model removed as well?

45. Can you place a projected view of a model on a different sheet from the original view?

46. List some of the advantages and disadvantages of using drawing templates, including some of the various options.

Project Exercises

Here are the last three parts for the cart. Pretty routine stuff!

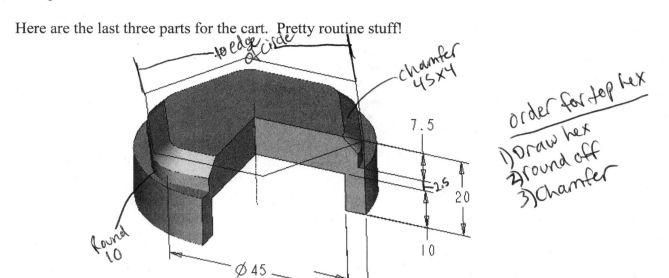

PART: *pillar_cap*

In the part below, make sure you use a pattern to create the four mounting holes - we will need that in the next lesson.

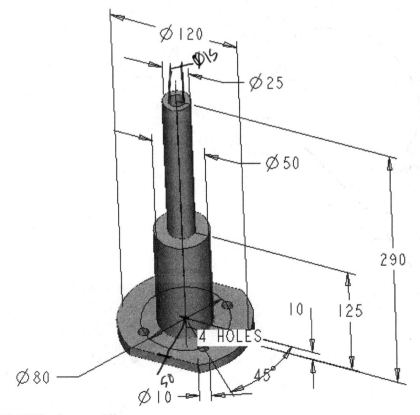

PART: *front_pillar*

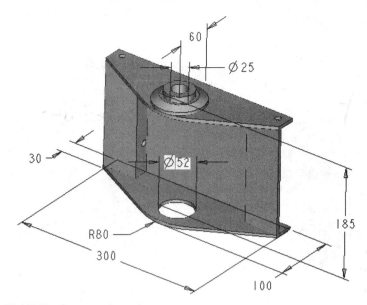

PART: *frame_front*

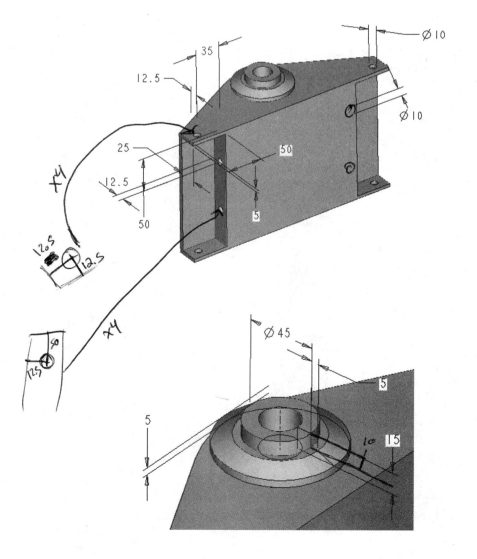

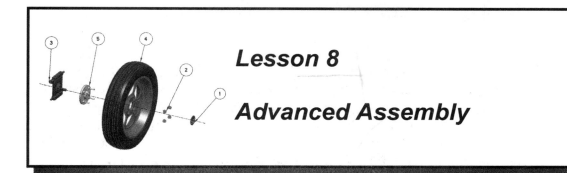

Lesson 8

Advanced Assembly

Synopsis

Creating parts using *Merge* and *Mirror*; advanced component assembly techniques (*Repeat*, *Reference Pattern*); using component interfaces; creating parts in the assembly; flexible components; using Pro/PROGRAM in an assembly; assembly shortcuts using Interfaces (drag-and-drop, *AutoPlace*); creating a drawing with BOM and balloons

Overview

In this final lesson, we are going to assemble all the components in the cart. The various parts have been described in the project exercises at the end of the previous lessons. First, we will examine the assembly plan to determine a strategy for assembly. The cart makes extensive use of subassemblies to help organize the work. We will utilize some useful functions to create new parts by *merging* (like welding them together) and *mirroring*. Then we will begin putting together the subassemblies. This will involve some advanced component placement functions for duplicated components, like *repeating* and *patterning*. Component *interfaces* will be introduced as a quick way to implement design intent and automatically place common components (i.e. that occur in multiple places in the assembly). Component placement by *drag-and-drop* and automatic placement will also be demonstrated. Defining the spring as a *flexible* component will allow the cart assembly to automatically adjust to changes in the specified ground clearance. Finally, we will create a couple of assembly drawings which will include a *bill of materials* and *balloons*.

Creating the Assembly

We are finally going to put together all the cart components. Along the way, we will explore a number of functions available in assembly mode. Assembly functions allow you to do a lot more than just constraining the components. We will also create some new parts by merging and mirroring other parts, set up relations, and make a program to control some aspects of the geometry of the assembly and individual components using simplified user input. We will, unfortunately, not have time here to go into the intricacies of mechanism connections that would allow various components in the assembly to move. Before we start any of that, since the cart is a pretty complicated system, it is necessary to develop a plan of attack - the assembly plan.

The Assembly Plan

Before we start creating an individual part, we (should) probably have a pretty good idea how we are going to select various features and the regeneration sequence. The same planning process is required before starting an assembly. How will we organize the various components into subassemblies? In what order should we put the components and subassemblies into the main assembly? How will we set up and manage the placement constraints and the parent/child references created? What additional assembly features like datums will be required, and where and when should these be created? Answering these questions before we start putting the assembly together will allow us to organize our efforts and prevent back-tracking later on. The end result is that we will be more efficient in our work and will have a cleaner and more flexible model.

The main difference between an assembly plan and a part plan is the use of subassemblies. There is no (direct) analog to subassemblies in part creation (although UDFs come close). We will make extensive use of subassemblies in the cart. The total assembly can be thought of as a tree structure, as shown in Figure 1. The major branches of the total assembly are subassemblies. This does not show individual components

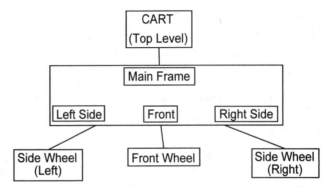

Figure 1 The cart assembly tree structure

added to the assembly at the top level (like all the bolts!) nor the individual components used to connect the subassemblies (like the wheel suspension arms). The physical configuration of these subassemblies is shown in Figure 2.

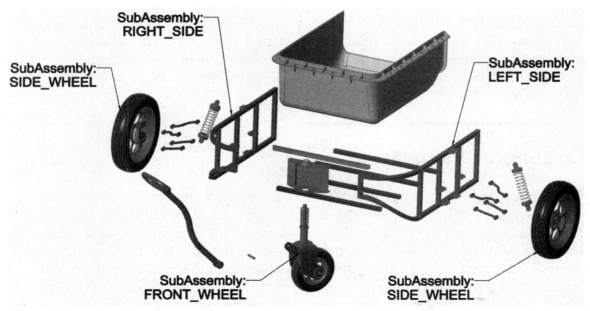

Figure 2 The major subassemblies in the cart model, plus single parts

Each of the subassemblies, of course, could contain additional subassemblies at a lower level. Once a subassembly is created, it is treated as a component in the parent. This nesting of subassemblies can go many levels deep. Once the assembly plan has been developed, we can start at the bottom of the tree to create the lowest level subassemblies.

Before we actually start putting things together, recall some basic aspects of working in assembly mode:

◆ **Assembly Constraints**. Components are located in the assembly using constraints to existing components and assembly features. A fully constrained component involves all six degrees of freedom of location and orientation. When you are placing a component (particularly involving alignment of axes) sometimes Creo reports that the component is fully constrained, yet it still technically has a rotational degree of freedom. This is often all right (for example when placing the bolts) but sometimes not (for example when adding the hubcap). If this occurs, you can just add an additional constraint.

◆ *3D dragger* - This is a tool that allows well-defined movement of a component in an assembly. The component can be translated or rotated using arrows or rings on the dragger, respectively. Free motion is possible by dragging the center. If you hold down the RMB on the center, you can align the dragger direction/orientation with existing geometry.

◆ *Place* vs *Package*. A component that is fully constrained in the assembly is "placed." If it is not fully constrained, it is "packaged." If the component is constrained in any way, then the movement allowed with the 3D dragger will automatically maintain the existing constraints. Packaged components are indicated with a special symbol in the model tree. Remember that children of a packaged component are also considered as packaged. This often happens when the base component is brought into the assembly without applying constraints.

◆ *Separate Window* vs *In Assembly*. When a new component is being assembled, you can view it either in a window by itself or in the same window as the assembly. This is a matter of personal preference. If the component and the assembly are very different in size, it is probably easier to use the separate window.

◆ **Mouse shortcuts**. While you are setting up the constraints on a new component, you can move the component in the assembly very easily using mouse shortcuts. This movement will be consistent with any currently defined constraints. For a totally unconstrained component the shortcuts are:

CTRL-ALT-left	slide normal to screen
CTRL-ALT-middle	rotate/spin relative to component spin center
CTRL-ALT-right	pan

Let's get started...

The Side Frames

We'll start by putting together the tubing to make the right side frame. You will note that we did not create any components for the frame on the left side of the cart. We will do that here using a mirror function. Also, we presume that the side frame will be welded together to form a single piece. So, we will create an assembly of the side frame and then 'weld' it together using a special merge procedure to create a new part. Do not confuse this with the use of all the welding tools in Creo. These tools allow you to specify full welding parameters (weld shape, rod type, weld parameters, and so) for the actual welds in an assembly. These welds can be identified along edges or surfaces in the assembly. See *Applications > Welding* some time to explore this capability. We will not do that here.

Create an assembly (units: mmNs) called *[frame_right]*. The assembly should have named default datum planes and named views for TOP, FRONT, RIGHT, and so on.

Assemble the lower right side tube, *fram_low_rgt.prt*. Constrain this to the assembly default datums using three *Align* constraints or just use the *Default* option in the constraint list in the dashboard (or in the RMB pop-up).

Retrieving a Family Instance Component

The next component to place is one of the 4 vertical tubes along the side of the cart. The tube we want is one of the instances in the family table of *tubing.prt*. Bring this into the assembly with

> *Assemble*
> *tubing.prt*
> *T25X325*

We are going to make a pattern of this component; therefore, we need a dimension to be incremented between instances of the pattern. We will use a *Distance* constraint for this (with a small offset, like 15). Assemble the tube as shown in Figure 3.

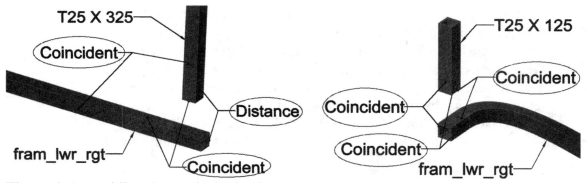

Figure 3 Assembling the vertical tube for a component pattern

Figure 4 Assembling the front vertical tube

Bring in another instance (the **T25X125**) and constrain it as shown in Figure 4.

Assemble the top frame part *fram_upp_rgt* by constraining it using *Coincident(Mate)* constraints to the top of the front vertical tube. (Why not the back one?)

Component Patterns

Now we can pattern the vertical tubes. Select the back vertical tube and in the RMB pop-up menu, select:

Pattern

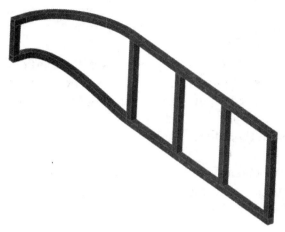

Click on the offset dimension. The pattern increment is **275**mm. There are **4** instances in the pattern. Once patterned, change the offset dimension of the pattern leader to **0** and *Regenerate* the part. The assembly should now look like Figure 5.

Save the assembly. We are not quite finished with it yet, so don't remove it.

Figure 5 The pattern of 4 vertical tubes

Merging Components

Note that the *frame_right.asm* contains individual components constrained to each other. Zoom in on one of the joints to see the edges formed by the individual components. As stated above, our intent for the cart is that these tubes will be welded together, making them a single component. We can do that in assembly mode by merging these tubes into a single solid. At the same time, we will create a new part file containing the merged tubes.

Create an empty part called *[frame_right]*. We have used the same name as the assembly. This is not necessary but will help us keep everything organized. Creo will not get confused between the part and assembly files of the same name.

Activate the window containing the assembly *frame_right*. Bring in (*Assemble*) the empty part *frame_right* (pick it from the *In Session* button in the Open dialog window) and constrain it to the assembly datum planes in the assembly *frame_right* or select *Default Constraint*. Observe the part location in the model tree.

Now we will perform the merge. In the **Component** group drop-down menu, select

Component Operations > Boolean Operation

The default operation is *Merge*. We have to identify two major items, in this order:

♦ the component(s) to be merged to (our empty part *frame_right*). This will be the *modified model*.
♦ the components to be added to the merge set (all the tubes). These are the one (or more) *modifying components*.

The first selected component is the one whose geometry will change as a result of the merge[1], in our case the empty part *frame_right*. Select that in the model tree now.

Now click in the collector for **Modifying Components** to select components that will be added to (i.e. merged with) the previous component. The easiest way to do this is to CTRL-click on each of the tubing components in the graphics window - there are seven in all.

Leave the rest of the settings in the **Boolean Operations** window as default. The important one (for our purposes now) is **Update Control(Automatic Update)**. Setting this means that any changes in the original components will be propagated automatically to the final merged part. For example, if we change the location/diameter of any holes or wall thickness of the tubing parts, those will also change in the "welded together" part.

We can also choose if we want to bring datums into the merged part. The default is not to do this.

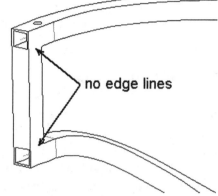

Come back later to explore these options. For now, select *OK > Done/Return* and go to the *frame_right* part window. Zoom in on one of the tubing joints - there are no lines showing between components. We have created one solid. See Figure 6.

Open the model tree for the new part to see how the merged components have been identified. Check out *Information >* **Figure 6** The merged right frame part
Feature Information in the RMB menu for one of the merged features.

Add some holes to the right side part according to the dimensions shown in Figures 7 and 8. Make the hole through the middle two vertical members (Figure 7) as a single hole going through both members. This provides a single axis that we will use to assemble the brackets a bit later in the lesson.

Save the part *frame_right*. Save the assembly file and remove it from your session.

Creating a Part using *Mirror*

A very useful tool for creating symmetric parts is the ***Mirror*** function. Although there is a mirror function for features in part mode, creating mirrored components is best done in an assembly. There are several variations of the mirror function. Also, you can use either the actual assembly containing the component and its mirror, or you can create the mirror copy in a dummy assembly that will be discarded later. In this lesson, we will use a dummy assembly file. Create one now using your default assembly template; call the assembly something like *[dummy]*. We will not be keeping this permanently.

---

[1] You can merge to a component that already has solid features.

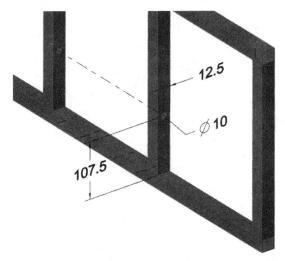

Figure 7 Bracket holes in vertical tubes

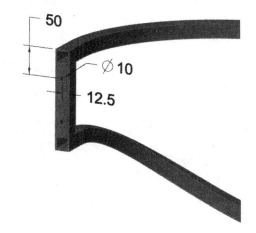

Figure 8 Holes for mounting the front frame (lower hole has same dimensions)

Assemble the merged part *frame_right* that we just created (make sure you get the part, not the assembly of the same name). You should be able to find this in session. Constrain it to the default datum planes in the new *dummy* assembly or use **Default Constraint** in the RMB pop-up.

Now we create the mirror copy. In the **Component** group, select

Mirror Component

In a mirror operation in assembly mode, there are two main options: setting the model type and dependency control. The model type can be set to one of the following:

(a)   Mirror the total part using geometry only, resulting in a new part with a single Merge feature in the model tree. This is the default setting. The dependency of the new part on the original is set using dependency control options described below, OR

(b)   Mirror all the individual features of the original, resulting in all the features appearing in the model tree in the new part. This is automatically set to be independent of the original part, OR

(c)   Placing a copy of the same part at a mirrored location within the assembly. No new part is actually created (despite what you may have entered in the previous dialog window) and therefore there is no question of dependency since it is actually the same part file placed at a new location, more like a simple *Copy* command.

The dependency control is used to determine if modifications to the original part will affect the mirror copy or not, and to determine if changes to the location of the original part in the assembly will affect the mirror copy.

You might like to come back later and try out some of these options. For now, select the options **Mirror(Geometry Only)** and **Geometry Dependent** (these are defaults), but turn off the **Placement Dependent** option.

Click in the **Component** collector and then pick the frame part on the screen or in the model tree. Now select the mirror plane - the vertical face of the short vertical tube at the front of the frame, as shown in Figure 9, will do. You can use any vertical plane parallel to this one. Enter a new name: *[frame_left]*. There is a **Preview** check box at the lower left corner of the dialog window. If the mirrored part appears as in Figure 9, select **OK**. The part mirrors in the assembly.

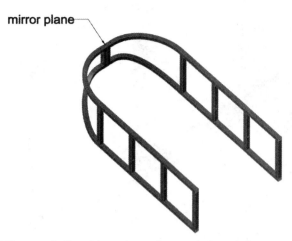

Figure 9 Creating the mirrored frame part

Note that the parts are not joined (as in merged). To prove that, **Open** the new part *frame_left* into a separate window by highlighting it and using the RMB pop-up menu. Have a look at the model tree for the new part - it contains a single, uneditable **Merge** feature. The new left frame part has not yet been saved to disk; do that now. You do not need to save the *dummy.asm*.

Examining Dependencies

Let's examine the relationship between the original tube parts and the new merged and mirrored left frame part. Open the part file for the upper horizontal frame tube (*fram_upp_rgt*) which should (if you've been following along closely!) have a pattern of holes. In the upper tube part file, modify the diameter value of the hole pattern leader (make it **20**) and **Regenerate** the part. Now go to the new part *frame_left* and **Regenerate**. Magic! The holes have changed. The change has propagated from the original tube *fram_upp_rgt.prt*, through the original assembly *frame_right.asm*, the merged part *frame_right.prt*, and into the mirrored part *frame_left.prt*. Go back to the upper tube part file and return the hole diameter to the original value (**10**). Note that the assembly *frame_right* and the merged part *frame_right* need to be loaded in session to do this, although they do not need to be displayed.

You can now make sure everything is saved, then remove all objects from your session.

The Side Frame Subassembly

The first subassembly of the cart is the side frame. These will include the side frame parts made above, and the brackets to hold the wheel suspension arms and spring. We will use a couple of handy tools for constraining identical components in a number of places on the assembly. We will also add some assembly features that will be used to assemble the side wheels and make their position adjustable.

Start by creating a new assembly called *[right_side]*. Bring in the merged right side part using

Assemble

and select *frame_right.prt*. Make sure you pick the part and not the assembly of the same name. Align the part with the assembly default datum frames.

Now we'll bring in the brackets for the wheel mounts. The placement constraints for these are very similar and we will use some special commands in assembly mode to expedite their placement.

Repeating Components

Bring in the bracket for the vertical frame members:

Assemble

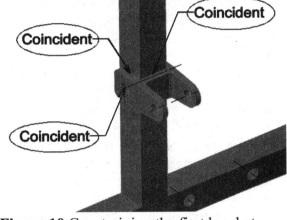

and select the part *arm_vbrack*. Set up the assembly constraints for this first bracket as shown in Figure 10. There are two *Coincident(Mate)* constraints and one *Coincident(Align)* (for the hole/axis).

The only difference between the placement of this component and the bracket on the next vertical tube is the assembly reference

Figure 10 Constraining the first bracket

for the *Coincident(Mate)* constraint on the side of the tube above the bolt hole - the *Mate* constraint on the front refers to the same surface at the other bracket location, and the

Align refers to the same axis. We can save a lot of mouse clicks by using a special function for repeating the placement of components. With the latest bracket highlighted, in the RMB pop-up menu, select

Repeat

This brings up the **Repeat Component** dialog window shown in Figure 11. In the **Variable Assembly Refs** pane is the list of constraints for the chosen component. Follow the prompts in the message window. Click on each of these to see the assembly references highlight on the model (wireframe works best for this). Highlight only the **Coincident(Mate)** constraint that involves the inside vertical face with the hole. This chooses the constraint to be repeated and changed for the next component. Then select

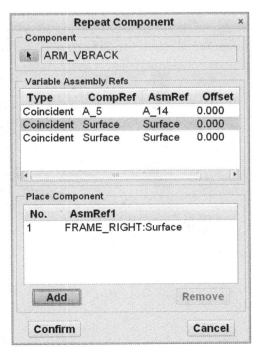

Figure 11 *Repeat* dialog window

Add

Follow the instructions in the message window: pick on the corresponding surface on the next vertical tube in the frame. See the constraint indicated at the top of Figure 12. A new bracket appears at the new location. Select *OK*.

It's that easy! *Repeat* works very well when only one or two references need to be modified for the repeated component.

Bring in the other bracket, *arm_brack*. We can place this with the constraints shown in Figure 12. Align the axes of one hole, mate the back surface of the bracket with the tube, then set the side surface of the bracket parallel with the top of the tube. Note that Creo will sometimes give you a message that the component is fully constrained, and when you accept the part it ends up upside down. The "fully constrained" message is generated when Creo thinks it can locate the part but it is using some internal rules to determine the orientation. Sometimes it guesses wrong, usually involving a degree of freedom of rotation around an axis. You can fix the problem by turning off the *Allow Assumptions* box in the **Placement** slide-up panel in the dashboard and then adding an additional **Parallel** or **Coincident** constraint.

Use the *Repeat* command to place duplicates of the *arm_brack* part at the other locations on the side frame shown in Figure 13. You can *Add* both of these at the same time. You actually only need a single new axis reference since both the previous alignment surfaces still apply. This is why the constraints of Figure 12 were not set up to use both hole axes, although that would be a clearer design intent. For each copy, you only need to select a new axis for the hole alignment.

We must also add these brackets to the left frame. We will do that shortly using another new tool in Creo that will make repetitive assembly tasks much easier - component interfaces. Before we leave the right frame, however, we will look at using datum curves in an assembly to drive the geometry - a simplified version of a skeleton model. Before we proceed, save the assembly.

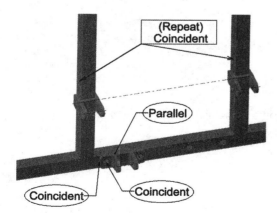

Figure 12 Assembling the lower bracket

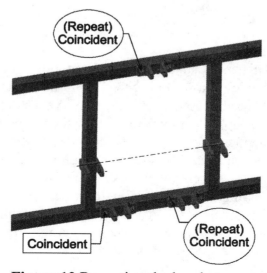

Figure 13 Repeating the bracket

Creating a Skeleton Feature

We're going to add some assembly features (a datum curve and some axes) to the side frame to help us out later by supplying references when we assemble the side wheel assembly. The problem is that, for the side wheels, the S-shaped side arms, the mounting plate, and the frame form a parallel 4-bar linkage. Consider the difficulty involved in constraining these components, since the four links must be assembled to form a closed chain. Suppose you start with the frame and try to assemble one of the side arms. Without the wheel mounting plate already in the assembly, there is nothing to assemble the other end of the side arm to. And the mounting plate can't be assembled first without the side arm to constrain to. Try this out some time! We'll get around this sort of vicious circle by creating some assembly features that will allow us to constrain the components properly[2] and also allow us to easily change the vertical position of the wheel using Pro/PROGRAM. These assembly features are typically datum planes and curves. It is possible to construct parts and subassemblies containing nothing but datum features. These are called *skeleton models*. Such models are then used in the creation of solid features, the geometry of which is driven associatively by the skeleton.

Start by creating a sketched curve in the right side frame assembly (see Figure 15). For the sketching plane, select the outside vertical plane of the bracket closest to the back of the assembly. For the sketching references required by Intent Manager, all you need to select are the axes of the two outer holes in the brackets. Note that these holes are 120mm apart (the height between the pins on the wheel mounting plate) - this will yield a parallel 4-bar linkage. The sketch for the curve is shown in Figure 14. Note that the length of each long edge in the curve corresponds to the dimensions of the

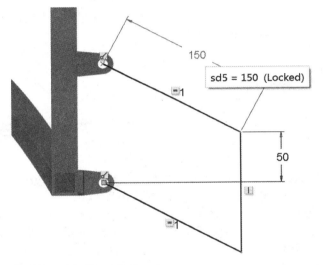

Figure 14 Sketch for datum curve

side arms. Also, note the location of the vertical height dimension. We will be using this later to control the position of the entire suspension system. You can see the effect this will have by *Lock*ing the arm length dimension (as in Figure 14) and dragging on one of the free vertices.

---

[2] Actually, since there are movable components here (i.e. a mechanism), the best way to set up this assembly would be to use mechanism connections (pin joints). We will not have a chance to discuss mechanism functions here. There are some useful on-line tutorials for these functions in the online *Help*.

Create a couple of datum axes through the corner vertices of the sketched curve and normal to the ASM_RIGHT datum plane. See Figure 15.

We are finished with the right side subassembly. Save the assembly and remove it from the session. We must now do this all over again on the left side to create another subassembly called *left_side*. Although we could use the assembly **Mirror** command to do that[3], we will create a new assembly for that and bring in the part *frame_left*, the five brackets, and then create the sketched curve and axes. This will give us a chance to have a look at another assembly function in Creo to help with repetitive component placement - component interfaces.

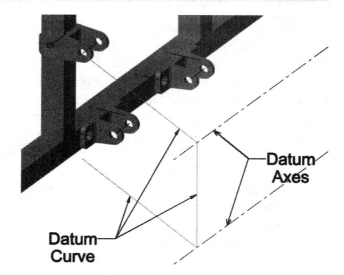

Figure 15 Datum curve and axes in side frame assembly

Component Interfaces

This function in Creo is directed at making assembly of multiple copies of identical components very quick and efficient. The idea is that for these components, the set of references and constraints (collectively called the *interface*) will be the same for every placement of the component in the assembly, so they might as well be set up ahead of time (and only once!). By defining the interface in the component itself, all you need to specify in the assembly are the matching assembly references. This cuts the number of mouse clicks in half each time the component is brought into the assembly. Furthermore, with the component interface defined, other methods of placing the component become possible including drag-and-drop and automatic placement. We will see both of these this lesson. It is also possible to permit assembly constraints only to designated component references (surfaces or axes, for example) in the interface. This leads to more standardized assembly procedures in large, multi-user projects. There are some restrictions on the types of constraints that can be used with component interfaces. Basically, if you stick to surfaces, datum planes, and axes, all will be well.

There are two ways of defining a component interface. The first is when the part is being created. The second is while the part is being assembled. We will see both methods here. There are also several ways of using the interface (manual, drag-and-drop, and automatic) which we will also see.

---

[3] Not a good idea here since that would create mirror copies of all the brackets too, which is not necessary and would just clutter up the model.

Creating a Component Interface in Part Mode

Open the part *arm_vbrack*. Recall the
constraints that were used on the right frame:
two surfaces were mated with the vertical tube,
and the hole axis was aligned with the hole in
the tube. See Figure 16. To create the interface,
starting in the **Model** ribbon, in the **Model
Intent** group select:

Component Interface

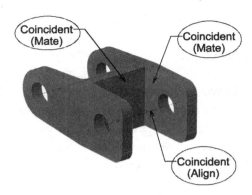

This opens the **Component Interface** window
(Figure 17). We can provide a name for the
interface (a default INTFC001 is given). It is
possible to define multiple interfaces for a

Figure 16 Selecting references to be
used in the component interface

single component, so having descriptive names is useful. Enter the name *[vbrack1]* then
click on **Automatic** in the collector on the left side. Pick on the axis of the hole (see the
Align constraint in Figure 16). By default, a **Coincident(Align)** constraint is set up by
default, exactly what we want. Now click on **New Constraint** and then select one of the
vertical surfaces that will mate with the outside of the tube. Repeat this for the other
vertical surface. Both will default to **Coincident(Mate)** constraints. The window should
now look like Figure 17. Observe that we could also specify offset values and other
variations of these constraints. Note that it is up to you to determine if these constraints
will fully constrain the component. Accept the dialog window and save the part. Have a
look at the bottom of the model tree for the part. This now contains a **Footer** feature that
shows the interface. See Figure 18.

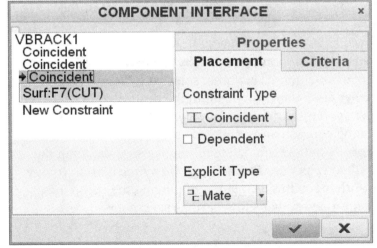

Figure 17 Creating a component interface definition

Figure 18 Model tree
showing the footer containing
the component interface

Now we can proceed with the left side subassembly. Create a new assembly called
[left_side]. Assemble the part component *frame_left* at the default location.

Now we will add the component *arm_vbrack* to the left side assembly. Select the *Assemble* command and select the part from in-session. The assembly dashboard opens but with a few additional tools we haven't seen before. On the top row, two buttons at the left are now active. The first button allows us to place a component using an interface. Since our bracket component now has an interface defined, this is the default. The second button allows us to place a component manually, as we have done up until now. There are two pull-down lists. The first is to **Interface to Geom** (default). The next list is for selecting one of the component interfaces defined in the part. In the **Placement** slide-up panel, the *vbrack1* interface constraints and component references are listed. All that is missing are the assembly references. The component display in the graphics window also shows the interface references and constraint types using flags. One of the references (probably the axis) is highlighted. As each constraint is highlighted, pick on the corresponding reference in the assembly. Observe the change in color of the constraint flags as the assembly references are completed. This is a semi-automatic way of setting up the assembly constraints. Accept the component.

There are a couple of easier ways of using an interface that will usually work for straight forward constraints. We will try these out by assembling the bracket to the other vertical tube.

Select the *Assemble* button again and select the component *arm_vbrack*. In the dashboard, pick the button for manual placement (second from left). Use the 3D dragger to drag the component closer to its final position and orientation near the desired mounting hole on the other vertical tube. Now select the *Place Using Interface* button. If the location is close enough, the component may snap onto the assembly, finding the necessary references to match those defined in the interface. If so, accept the component placement. If not, select the assembly constraint references as usual. Try double-clicking on the constraint flag. This opens a pop-up version of the Placement panel, showing component and assembly references. Go ahead and complete the placement of the bracket.

To try another variation of this drag-and-drop procedure, delete the bracket and launch the command again to bring it into the assembly. This time, select the *Place Using Interface* button immediately and then drag the component to the desired location. You might be able to get it there right away, but the component may also get "hung-up" on other references in the assembly as the component is dragged across the screen. If that happens, just select the *Place Manually* button and delete all the constraints. Drag the component to the desired location, then select the *Place Using Interface* button. It may take some practice to become comfortable with this method of placing the component, particularly in reading the subtle color queues that show various references.

To experiment with yet another method of using the interface, delete the component from the assembly again. Launch the *Assemble* command and select the *arm_vbrack* component. Now, in the dashboard, select the *Auto Place* button. A new dialog window opens, the **Auto Place** window. See the message window. All you have to do is click on the screen somewhere in the vicinity of where you want the component to be placed on the vertical tube. You can click on either the existing frame or even the empty space near the desired location. In the **Auto Place** dialog window, a number of possible locations are listed (as **Location 1**, **Location 2**, and so on). The default is to locate a maximum of five possible locations. If you click on one of the listed choices, the component will move to that location. Since this part has a couple of planes of symmetry, the change in orientation may be hard to detect (check the datum planes). When the placement that you want is obtained, select the double right arrow to move the selection to the right pane, then select *Close*. The component is placed and fully constrained. Middle click.

We will now bring in the other bracket component that doesn't yet have an interface defined. We will create the interface here in assembly mode (the previous bracket was done in part mode).

Creating a Component Interface in Assembly Mode

In the preceding part, we defined the component interface in part mode. There is another way to do this. Bring in the part *arm_brack* and constrain it manually to the lower frame member using the constraints shown in Figure 19. Do not leave the assembly dashboard just yet.

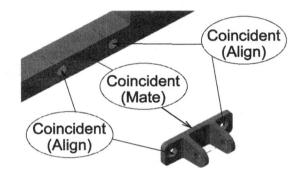

Figure 19 Defining the component interface for the horizontal bracket

These constraints are a bit different from the ones we used before so that the geometry will be more amenable to automatic placement of the component: the presence of two parallel axes a certain distance apart and perpendicular to the same reference surface doesn't happen very often in the assembly - when it does, the most likely candidate component to put there is the bracket. This might prevent some of the incorrect automatic placement behavior we saw earlier.

With the component fully constrained using a manual method, open the RMB pop-up menu in the graphics window and select *Save as Interface*. This opens a dialog window where we can name the new interface. Call it *[hbrack]*. Select *OK* and accept the component placement. Open the model tree and turn on display of all features. At the bottom of the bracket model tree you will see a footer feature with the new interface listed there.

Now we can go ahead and assemble more brackets. We could do this the same way as the previous bracket, perhaps using *Auto Place*. BUT... there is a really neat function available in Creo.

Open the Browser window, and use the Navigator to show your current working directory. Position the assembly so that you can see the target locations for the brackets, as in Figure 20. Select the part *arm_brack* in the directory listing. Then you can drag-and-drop it directly onto the assembly. The drop location should be reasonably close to the desired location (perhaps between the two holes in the horizontal tubes). Creo will look (in the vicinity of the drop point) for references in the assembly that match the component interface, and snap the component there. The component is fully constrained. Place both horizontal brackets this way.

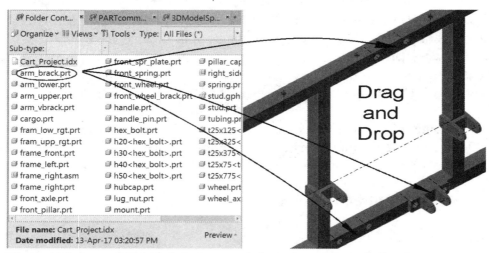

Figure 20 Assembling using drag-and-drop (requires a defined component interface)

In the left side subassembly, create the same datum curve and axes that we did in the right subassembly. Then save the subassembly and remove it from session. We will get another chance to play with component interfaces a bit later - including placing several copies of a part simultaneously.

The Front Wheel Subassembly

We'll use the front wheel subassembly to illustrate another easy way of quickly adding a number of related components to an existing pattern of features. These are the bolts connecting the vertical pillar to the front wheel bracket. The parts involved in the subassembly are shown in Figure 21.

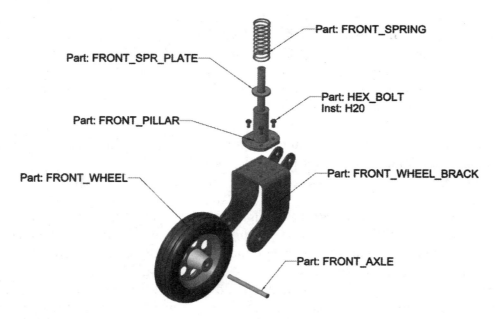

Figure 21 The *front_wheel* subassembly

Using your default template, start a new assembly called *[front_wheel]*. Again, Creo doesn't mind if you already have a part of that name since the file extensions are different. We'll start at the bottom of the assembly and work our way up. Bring in the part *front_wheel* and align it to the default datum planes. Then assemble the part *front_wheel_brack*. Use *Coincident(Align)* for the hole axes and the symmetry datum plane (you do have one?) of the bracket to the wheel. You may have to turn off assumptions (in the Placement panel) and add a *Parallel* constraint to make sure the bracket is vertical. Next, bring in the part *front_axle* and constrain it to the wheel (*Align* axes and symmetry plane). The assembly at this point should look like Figure 22.

Figure 22 Front wheel subassembly in progress

Figure 23 Identify the hole pattern leader on the *front_pillar* part

Assemble the part *front_pillar* to the top surface of the bracket. You can use *Coincident(Mate)* with the relevant surfaces and align the axes of two of the bolt holes. Observe the orientation of the flats on the pillar base. Make a note of which hole in the pillar is the leader for the radial hole pattern, possibly the one shown in Figure 23.

Using a *Reference Pattern*

The next part to bring in is the instance **H20** in the part *hex_bolt*. Assemble this using a *Coincident(Mate)* of the underside of the bolt head with the pillar surface. Then use *Coincident(Align)* for the bolt axis and the axis of the hole pattern leader in the pillar. Make sure you pick this axis and not the axis in the hole in the bracket (easier to do if you *Hide* the bracket). It is critical that all the bolt constraint references are to the same feature in the radial hole pattern. You can *Allow Assumptions* for the component placement. You may have thought of using a component interface for the bolt, which would certainly make the assembly easier. We will define an interface for the bolt a bit later.

Placing the remaining bolts is very slick. In the model tree or graphics window, highlight the bolt and in the mini toolbar select

Pattern

In the dashboard, the pattern type **Reference** should already be selected, so all you have to do is middle click to accept the pattern. A duplicate bolt will appear in each of the four holes in the radial pattern. Thus, we see that a reference pattern of components is created exactly the same way as a reference pattern of features.

Bring in the final two parts for the front assembly. The first is the part *front_spr_plate*, whose lower surface **Mate**s to the shoulder on the pillar about half way up and whose axis **Align**s with the pillar axis. Finally, the part *front_spring* can be assembled by aligning its axis with the pillar and using a **Mate** on the flat end of the spring to the top of the spring plate. The completed assembly is shown in Figure 24.

Using **Measure**, determine the distance from the top flat surface of the spring to a parallel plane tangent to the bottom of the wheel (you might like to create a datum plane there). We will need this distance later when we do the total assembly. It should be about 620mm, depending on how you made the tread pattern on the wheel.

622mm

Save the assembly and remove it and all its components from the session.

Figure 24 Completed *front_wheel* subassembly

The Side Wheel Subassembly

Start up a new assembly called *[side_wheel]*. The components are shown in Figure 25. Assembly of these components is going to be pretty routine. We can use another **Reference Pattern** to place the lug nuts on the wheel if you constrain the first one to the hole pattern leader.

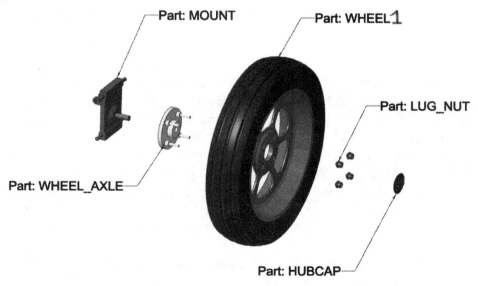

Figure 25 The *side_wheel* subassembly

To make it easier to locate the relevant assembly references, bring in the wheel first, then assemble the part *lug_nut* to the hole pattern leader. The lug nut is placed using a *Coincident(Align)* with its axis, setting the conical surface *Tangent* to the chamfered pocket on the wheel, and allowing assumptions. You can than use *Reference Pattern* to place the additional lug nuts. For all the remaining components in the assembly use *Coincident(Align)* to the axis of the wheel. The part *wheel_axle* is placed using a *Mate* constraint between the shoulder on one of the mounting studs and the inside surface of the wheel (the side opposite the lug nuts). You will also need to *Align* one of the studs with one of the holes in the wheel. The studs should just

Figure 26 Finished *side_wheel* assembly

protrude through the lug nuts. The part *mount* mates with the back surface of the wheel axle part. You may need an *Orient* constraint on this to get the proper orientation of the mounting plate. The outer lip on the hub cap mates with the outer surface of the wheel and aligns with its axis. You may have to use an *Orient* constraint on the hub cap if you have placed a design on the outer surface and you want it the right way up! The finished assembly is shown in Figure 26.

IMPORTANT: You might like to create a named exploded view of this assembly (as in Figure 25 - without the labels!) for the drawing exercise at the end of the lesson.

Save the assembly and remove it from the session. We will use an identical assembly on both sides of the cart, so you don't need to duplicate this.

The Frame Subassembly

The frame structure is our last subassembly. There is not anything new here, although you will get to use a component pattern again.

Create a new assembly called *[main_frame]*. The components are shown in Figure 27. You might like to rename the default datums (MAIN_RIGHT, MAIN_TOP, MAIN_FRONT) since things are going to get a little crowded soon. Start by bringing in the part *frame_front* and aligning it to the default assembly datums. Bring in the subassemblies *right_side.asm* and *left_side.asm* and mate them to the surfaces of the pockets on the back of the *frame_front*. All the holes should line up! Note that we could have created the mirrored part *frame_left* in this assembly.

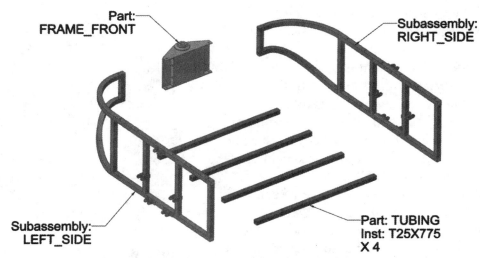

Figure 27 Exploded view of the *main_frame* assembly

Now assemble the rear cross-frame member using the instance **T25X775** in the generic part *tubing*. Since we are going to pattern this, use a ***Coincident(Distance)*** dimension to the vertical back face on the right side subassembly. Any offset value will do, but note it may have to be entered as a negative value (try **-50**). The end of the tube can ***Mate*** with the inside surface of the right side subassembly, and the top of the tube can ***Align*** with the upper surface of the lower horizontal frame member. When the tube is placed, the frame should look something like Figure 28. The tube should just reach across to exactly meet the frame on the left side.

Pattern the tube by highlighting it and in the RMB pop-up menu selecting

Pattern

For the patterned dimension, select the offset value. Enter the increment of **275**, and create **4** instances in the pattern. Accept the pattern and ***Edit*** the offset dimension to **0** so that the pattern leader is flush with the back end of the frame. This completes the frame, so save it and remove all objects from your session.

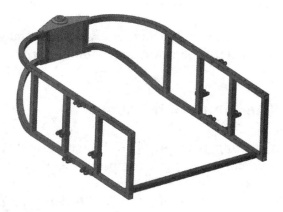

Figure 28 First cross member placed for patterning

The Main Assembly

It's time to put the whole cart together. This involves putting together the subassemblies made previously and a few individual parts. We will use some additional assembly datums to help us position various components and subassemblies. One part is missing (can you figure out which one?) and we will create it in assembly mode. We will also introduce the use of a flexible component (the spring on the side wheels) to allow its length to change automatically when the assembly is modified by changing the ground clearance using an input parameter.

Start up a new assembly called *[cart_total]*. Rename the default datums (CART_RIGHT, CART_TOP, CART_FRONT). The first component is the subassembly *main_frame*. Do not use the default placement. Instead, *Align* the FRONT and RIGHT assembly default datums in *main_frame* and *cart_total*. You will probably have to use preselection to pick these out of the crowd (cycle through the choices with the right mouse button), or use the separate window for the incoming component. We want the frame to "float" above the cart TOP datum so that when the front wheel assembly is added, the wheel will just touch it. The third constraint is therefore a *Distance* from the TOP datum in *cart_total* to the underside of the top plate in the *frame_front* part. The offset distance is the height of the *front_wheel* subassembly (from the bottom of the wheel to the top of the spring) that we measured previously. This dimension should be around 620mm. Don't worry if this number is not exact - getting the wheel to exactly touch the TOP datum is not critical.

Adding the Suspension Arms

Now we can bring in the side arms for the wheel suspension. Start on the right side and bring in the *arm_lower* part. The two lower arms face in opposite directions so that the separation between the arms at the outboard end is less than the separation at the cart frame end. Start by aligning the axes in the bosses at each end of the arm. *Align* one end with the forward bracket axis and the other end with the lower datum axis. See Figure 29. The final constraint is a *Mate Offset* between the flat surface of the appropriate boss and the inside face of the bracket. The offset distance is **0.5**mm.

The next part, *arm_upper,* is assembled in the same way to the forward upper bracket. This might be a good time to try to set up a component interface, since two more arms are required on the other side of the frame.

The second lower and upper arms can be easily placed using the *Repeat* command. For each arm, all you have to do is pick a new assembly reference surface for the *Coincident(Offset)*, since the other axes coincide for each part. The arms will automatically flip over if you pick the correct surface on the brackets. Or, of course, you could define component interfaces for these arms (since you have to place four of them, this effort might be worth it).

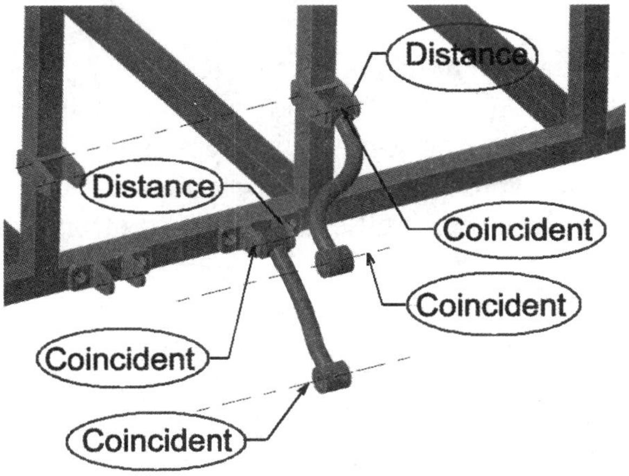

Figure 29 Assembling the suspension arms

Just for fun, click on the suspension datum curve and change the height dimension to the upper vertex to 100 and *Regenerate > Automatic*. All four side arms should rotate with the new datum frame position. Put the height dimension back to **50** before proceeding.

Adding the Side Wheel Subassembly

Now bring in the *side_wheel* subassembly. Place this using two axis alignment constraints between the assembly datum axes and the axes of the upper and lower pins on the mount plate. Finally, *Mate* the forward side face of the mounting plate with the inside face of the lower suspension arm. See Figure 30.

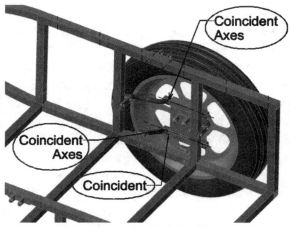

Figure 30 Assembling the *side_wheel*

The side arms and the side wheel subassembly now need to be added to the left side of the frame. Your system performance will probably improve if you suppress these components on the right side. As we did above, add the side arms first, then the wheel subassembly. Remember that you can use *In Session* in the **Open** dialog window when selecting components to be brought into the assembly.

There is a small dimensional error in the suspension system. To see this, view the assembly from the side and zoom in to examine the clearance between the four side arm bosses and the wheel mounting plate. These are not all the same. Can you figure out the problem? Can you fix it? You will have to move one component and change one assembly constraint. This is not critical to what follows.

Now is a good time to save your *cart_total* assembly.

Adding the Front Wheel Subassembly and Parts

The next component to add is the *front_wheel* subassembly. Before starting that, suppress (or *Hide*) the side wheel assemblies and arms. To make the position of the front wheel adjustable, we will first create a new assembly datum plane at an angle to the CART_RIGHT. The angle will be adjustable (we'll do that later with Pro/PROGRAM). We'll align the vertical datum of the front wheel assembly to this adjustable datum plane in *cart_total*.

To create this datum plane in *cart_total*, select the **Datum Plane** toolbar icon, then pick the vertical axis through the hole in the *front_frame* part. The reference type should be **Through**. Then, CTRL-pick on the CART_RIGHT datum. Check the direction of rotation and enter an angle value of **30**. This should create a datum **ADTM1** in the top assembly.

Now bring in the subassembly *front_wheel*. This is constrained using **Coincident(Align)** between the symmetry datum in the subassembly to the new datum plane ADTM1 created above. Use **Coincident(Align)** of the pillar axis with the hole axis in the *front_frame* part. Finally, use **Coincident(Mate)** for the top of the spring with the underside of the top plate in *front_frame*. See Figure 31.

Figure 31 Subassembly *front_wheel* added

Figure 32 Adding individual parts to the cart - handle, pin, cap

The next part to add is the handle. We also want to make the position of this adjustable by controlling the angle from a horizontal plane. Create another assembly datum through the axis of the horizontal hole on the front of the *front_brack* and at an angle of 45° above the CART_TOP datum plane. Bring in the handle and place it using three **Align** constraints: the hole axis on the handle boss with the hole axis in *front_brack*, the SIDE datum of the handle with the symmetry datum created for the *front_wheel_brack* part, the TOP handle datum with the assembly datum created here at 45° above horizontal. You might need to use **Flip** here on the axis constraint.

Now bring in the part *handle_pin* and assemble it on the handle axis. If you created this as a both sides protrusion, you can align the symmetry plane with the appropriate plane in the *front_wheel_brack*. Otherwise, you will have to use an **Offset** with a vertical plane in *front_wheel_brack*.

The last part on the front assembly is the *pillar_cap*. Bring this in and use **Mate** and **Align** to assemble it to the top surface and hole axis, respectively, of the *front_frame*. See Figure 32.

Just in case disaster strikes, save the assembly now.

Adding the Cargo Bin

The last major component is the cargo bin. Suppress (or Hide) all the other components except the frame. Bring the part *cargo* into the session. **Mate** the underside of the sweep that goes around the top of the bin with the top surface of the frame. **Align** the hole axes of the rear holes on the left and right sides of the bin with the corresponding holes on the frame. See Figure 33. Check to make sure that all the other holes in the bin and the frame are lined up.

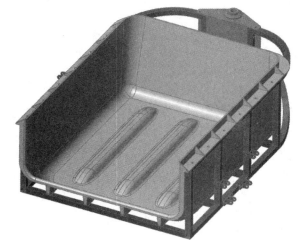

Figure 33 Assembly with the cargo bin

Creating a Part in Assembly Mode

Have you figured out which part is missing? We need a tubing frame member under the front edge of the cargo bin extending from one top side frame to the other. This was not made previously due to the uncertainty in the length, which depends on where the cargo bin is (and its size) and the curvature of the side frame members. The easiest way to create this part has been to wait until the assembly was together and then make the part to suit the geometry. We will create the tube as a both sides protrusion from the CART_RIGHT datum of the assembly *cart_total*.

In the **Component** group, select

> *Create*
> *Part | Solid*

Enter the name *[cross_tube]*. In the next dialog window, check the button

> *Create Features | OK*

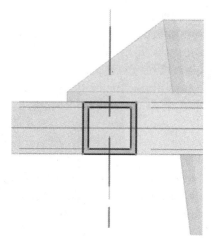

The component has been created (it has been added to the assembly model tree) and is now active (as indicated by the message on the screen and the symbol in the model tree). We are looking at the usual feature creation menu for parts. Select the *Extrude* tool and use the RMB pop-up menu to create an internal sketch. Pick the CART-RIGHT datum for the sketching plane and CART-TOP as the Top reference.

Figure 34 Sketch for *cross_tube* (constraints not displayed)

In Sketcher, pick the following references: the top, bottom, and interior surfaces of the side frame, and the axis of the hole in the cargo bin. We'll center the sketch on this hole. To help position this, sketch a centerline on the vertical reference (this allows us to use a symmetry constraint in the sketch). Then sketch two rectangles using the existing references. You should not need any dimensions on this sketch (Figure 34) if you remember the tube is square. Note that sketcher constraints are turned off in this image. When the sketch is completed, select *OK* and in the dashboard open the **Options** slide-up panel. Specify a *To Selected* depth in each direction (Side 1 and Side 2) and pick on the inside surface of the right and left top side frames. Accept the feature.

We need to make some holes in the *cross_tube* that line up with the holes in the cargo bin. Realizing that the cargo bin holes are a pattern, we will use a *Reference Pattern* to place the holes in the tube. To do this, we need to create the first hole at the location of the pattern leader in the cargo bin - find that location now. Note that the reference pattern leader is in another component. With the *cross_tube* part still active, select the *Hole* tool.

Make the hole with diameter *10*, and
depth *Through All*. For the primary
reference pick the axis of the pattern
leader hole in the cargo bin. This
automatically makes a *Coaxial* hole.
Ctrl-click on the top surface of the tube
(use pre-selection to make sure you get
this). Accept the hole.

With the hole still highlighted, open the
mini toolbar and select *Pattern.* Then
middle click. Presto! All the holes in the
tube are created. See Figure 35.

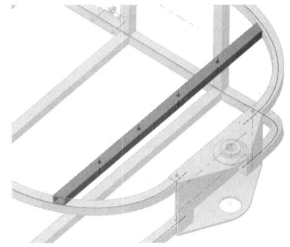

Figure 35 Cross tube member created

Since the cross tube is complete, we
need to make the top level assembly
active again. Select it at the top of the model tree; in the mini toolbar select *Activate*. The
notation disappears from the screen and the icon beside the cross tube part in the model
tree reverts to the normal one.

Setting a Display State

Have another look at Figure 35. Realizing that the cross-tube is a child of the cargo bin,
which appears to not be in the assembly, how was this figure produced? Try to suppress
the cargo bin to see what happens. The features of the *cross_tube* part are external
children of the cargo bin. With these identified, you could (don't do this now!) get the
desired view by *Suspending* the cross_tube. This keeps the tube visible, but it might
require special attention if the assembly ever needed regeneration. There are a couple of
ways to produce the display shown in Figure 35.

The first you may have tried already - selecting the cargo bin in the model tree and in the
mini toolbar, select *Hide*. This turns off the display without removing the part from the
regeneration sequence (and disrupting its children). Hiding works well for individual
components. Notice the change in the model tree component icon. To get the display
turned back on, select *Unhide* (or *Unhide All*).

If you want to control the display of many components at once or want to have a number
of pre-defined view states, select the *View Manager* tool in the Graphics toolbar. Select
the *Style* tab and *New* button. Enter a name for the style state *[no_cargo_bin]* and press
Enter (or select *Edit > Redefine*). In the resulting Edit window, select the *Blank* tab.
Over in the model tree, select the cargo bin part (and any others you want to get rid of) -
a notation **Blank** appears in the model tree. In the **Edit** window, select *OK*.

You can switch between the **no_cargo_bin** style state and the default state very easily in
the **View Manager**. Select the desired style state (like *Master Style*) and in the RMB
menu pick *Activate* to return to the default.

Create another view style called *[no_wheels]* that will blank both side wheels and the
front wheel. Check out the model tree. The component icons do not indicate the blanked

status of these wheels. However, you can add another column in the model tree that will do that (*Settings > Tree Columns > Type(Display Styles)*).

While you are in the Edit dialog for **no_wheels**, select the *Show* tab, pick the **No Hidden** option, then click on the cargo bin. Accept the definition and make **no_wheels** active. Thus, within a display style, you can set the display state of individual components.

Return to the **Master Style** set in the View Manager.

Assembling the Springs

We now want to add the springs on the side wheel suspensions. Resume the right side wheel subassembly. *Hide* the cargo bin, the two side wheel parts, and suppress the front wheel assembly.

Bring in the part *spring*. We'll assemble it to the right side wheel subassembly. Constrain the spring as follows (these will occur automatically): *Align* the horizontal axis at one end of the spring to the axis in the top bracket on the right side of the assembly, *Orient* the horizontal axis at the other end of the spring to the axis of the bracket on the back of the wheel mounting plate. Finally, *Align* the spring symmetry datum plane with the symmetry plane of the top bracket. If you look very carefully, you will see that the spring is not exactly the correct length to match the holes in the two brackets - it is too long by several millimeters - causing a large interference with the wheel mount. In the next section, we will fix this problem by declaring the spring to be a flexible component.

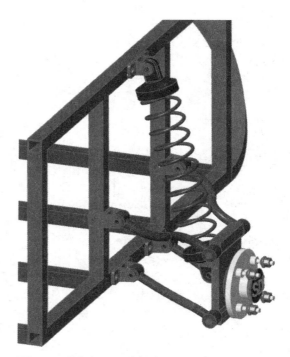

Figure 36 Assembled *spring*

Assemble another spring between the upper bracket and the wheel mounting plate on the left side of the cart.

Save the assembly.

Defining a Flexible Component

We want to set up a method to automatically compute the length of the spring in the suspension system for any vertical position of the wheel. Recall that the wheel position is determined by a dimension in the datum curve in the side frame subassembly (our skeleton). When the suspension is repositioned by modifying this dimension, we want to automatically update the spring length. We will implement this by declaring a flexible dimension in the spring component. The value to be assigned to this dimension is determined by the distance between the axes on the wheel mount and the top frame bracket where the spring is attached.

Highlight the spring part on the right side of the cart. In the RMB menu select *Make Flexible*. This opens a new dialog window (Figure 38) where we can define one or more variable dimensions for the part. Select the *Add* button ("+") and pick on the datum curve that defines the overall length of the spring (see the figures at the end of Lesson 3). Select the dimension that appears (see Figure 37) and middle click. It is added to the collector (Figure 38).

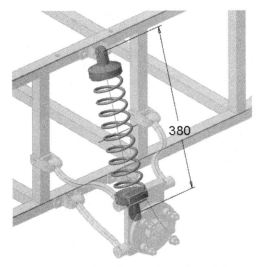

Figure 37 The dimension that drives the overall spring length

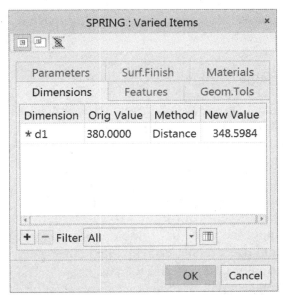

Figure 38 Defining a variable dimension for a flexible component

In the dialog window, open the **Method** pull-down list (currently *By Value*), and select the option *Distance*. You should now be in the **Distance** window where we will define how to determine the distance. All we need to do is select the axes in the top bracket and the wheel mount. The distance between these should be computed automatically (something like 348 mm). Accept the measure. Back in the SPRING window, select *OK*. The spring is then automatically regenerated and fits perfectly between the two axes. The number of coils in the spring should not change, if you have set up the part with a relation to control the helical sweep pitch based on the specified number of coils. We are in the component assembly dashboard. Middle click to accept the component. Observe the icon in the model tree that indicates a flexible component.

Repeat this procedure for the spring on the left side.

Select the right side datum curve for the suspension, and change the height from 50 to 100. *Regenerate*. Notice that the length of the spring on the left side does not change. Evidently, the flexibility applies only to the component instance for which it is defined.

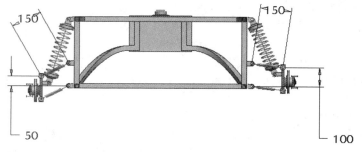

Figure 39 Two instances of spring with different lengths

Multiple instances of the same component can therefore have different flexible values in the same assembly. Multiple springs could all have different lengths. This also means that the component file on disk does not have the "flexible" value. In this respect, it is more like the generic part in a family table, with the flexible dimensions used to create individual instances stored in the assembly.

Using Pro/PROGRAM

Let's set up a procedure so that we can easily adjust the suspension height based on a desired ground clearance. This will use Pro/PROGRAM, a function introduced in an earlier lesson to input the desired clearance, and a relation. Referring to Figure 40, we will use Pro/PROGRAM to input the value of H (the ground clearance) and use a relation to calculate the dimension D used in the suspension. The relation will use the constant values for dimension A and B (and the bracket size).

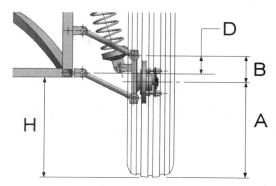

Figure 40 Dimensions required for ground clearance relation

Use the *Analysis > Measure* command to find out the values of dimensions A and B shown in Figure 40. These will be something like 282 mm and 75 mm, respectively. Referring to the figure, these are related by

$$H + 12.5 + D = A + B$$

Therefore, the value we want for the suspension datum curve is (use your values for A and B):

$$D = A + B - 12.5 - H = 345.5 - H$$

Using *Switch Dimensions* find out the symbolic name for the dimension D shown in Figure 40, something like *d6:5* (dimension 6 of component 5 - this is the height dimension in the skeleton curve). Then open Pro/PROGRAM and create the INPUT and RELATIONS statements shown in Figure 41. The INPUT area defines the ground clearance variable, and the RELATIONS area uses this value to change the datum curve dimension.

When you leave your editor, incorporate the design changes into the model and use **Enter**. Set a ground clearance of 250 mm.

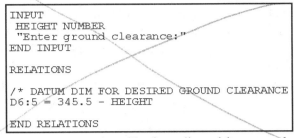

```
INPUT
 HEIGHT NUMBER
 "Enter ground clearance:"
END INPUT

RELATIONS

/* DATUM DIM FOR DESIRED GROUND CLEARANCE
D6:5 = 345.5 - HEIGHT

END RELATIONS
```

Figure 41 Design file for adjustable ground clearance.

We'll add a couple more input parameters to control the angle of the front wheel assembly and the handle. Open the model tree and select (with CTRL) the two datums ADTM1 and ADTM2. In the RMB pop-up, select **Edit**. The angle dimensions should show on the model, as in Figure 42. Use **Switch Dimensions** if necessary to get their symbolic names. The turn angle shown in the figure is *d11:1* and the handle angle is *d15:1*.

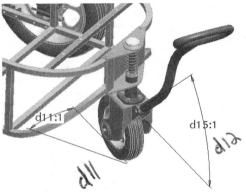

Figure 42 Assembly datum dimensions

```
INPUT
 HEIGHT NUMBER
 "Enter ground clearance:"
 TURN NUMBER
 "Enter turn angle (positive right):"
 HANDLE NUMBER
 "Enter handle angle (positive up):"
END INPUT

RELATIONS

/* DATUM DIM FOR DESIRED GROUND CLEARANCE - RIGHT SIDE
D6:5 = 345.5 - HEIGHT

/* TURN ANGLE FOR FRONT WHEEL - $ allows negative
$D11:1 = TURN

/* HANDLE ANGLE ABOVE HORIZONTAL - $ allows negative
$D15:1 = HANDLE

END RELATIONS
```

Figure 43 PROGRAM design file final version

Go to **Program > Edit Design** and modify the design file as shown in Figure 43. This involves adding two new input parameters for the turn and handle angles and two relations to assign these values to the appropriate parameter. Leave the design editor and incorporate the new design. Use **Enter** to set new values for the parameters, say **60** for the turn angle and **30** for the handle angle. Note that preceding a symbol with "$" allows you to enter negative values.

You may want to come back later and add another relation to control the datum curve dimension on the left side and define the left side spring as flexible too, otherwise the cart would not be level. Now is a good time to save the assembly.

Assembling Components using Automatic Placement

We only have 50 more parts to add to the cart - the bolts! **Resume** (or **Unhide**) the cargo bin, and **Suppress** all the wheel subassemblies (or use a **Style** if you have set one up). Adding the bolts is not a big job. Some useful tools are the **Repeat** command and the use of reference patterns. In this section, we will again use automatic placement of components. The bolts are all instances of the generic part *hex_bolt*. We will use the part

H50 on the suspension brackets (see Figure 44) and the part **H40** for all holes in the cargo bin and the front frame (see Figure 45). On the cargo bin, we will make good use of reference patterns. Go ahead and do the cargo bin bolts now - remember that to use *Reference Pattern*, the first bolt must be constrained entirely to the pattern leader hole in the cargo bin. When all the bolts have been placed in the cargo bin (there should be 14), *Suppress* the cargo bin and all its children.

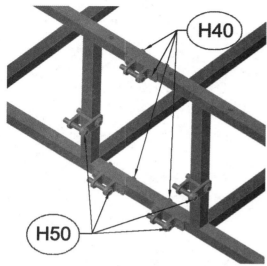

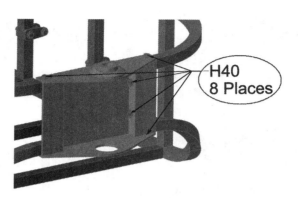

Figure 45 Adding bolts to *frame_front*

Figure 44 Adding bolts to brackets

Now for something new. In order to perform automatic assembly, we must first define a component interface for the bolt. Open the generic part *hex_bolt* and create a component interface (*Component Interface*) consisting of a *Coincident(Mate)* on the underside of the head, and a *Coincident(Align)* on the bolt axis. Name the interface something like FLUSH_BOLT. Save the part and return to the cart assembly.

Position the assembly so that you have a clear view of the side arm brackets on the frame. Select the *Assemble* toolbar button and choose the instance **H40**. Select the interface FLUSH_BOLT. Now select *AutoPlace* in the dashboard. Pick on one of the brackets near a hole where we want the bolt to go through a frame member (see Figure 44). The **Auto Place** window will open (Figure 46) showing a number (default 5) of possible placements of the bolt on the bracket nearest to the pick point and using the defined component interface. Picking them in the **Auto Place** window will display each placement on the model. When

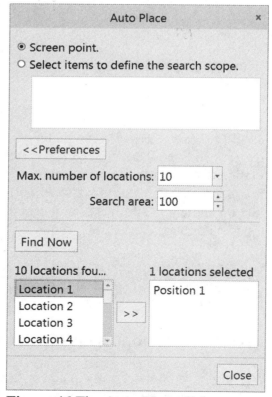

Figure 46 The **Auto Place** dialog window

you are happy with the choice, pick the double arrow button to move the location to the right pane. It may be that more than one location is what you want (there are two bolts in each bracket), so check them all. When you have them both, select *Close*. Do not leave the assembly dashboard just yet.

Open the **Placement** panel and select **New Location**, then pick the *AutoPlace* button again and pick on the model somewhere near the location for the next bolt. When the desired location is found, move it to the right pane in the **AutoPlace** dialog (Figure 46) and *Close* the window. You can continue doing this as often as you want to place multiple bolts in different locations on the model, without leaving the dashboard. Check out the commands available on the RMB pop-up.

If it seems like **AutoPlace** is not recognizing possible locations, check the **Options** tab in the **Component Placement** dashboard. The default is to not allow a component placement that produces an interference. This might occur with the H40 bolts on the front of the frame. Turn off these checks to place the bolts, and come back later to deal with this interference!

Also note the *Placement* button in the **AutoPlace** window. This lets you change the number of locations found and the size of the placement search area. Again, try these options out.

Another thing to try is the following: When assembling some H50 bolts, with **Max Position** set to the default, select on the side of one of the brackets on the vertical tube members. Several possible positions will be listed. Two of these are valid - select them both (with CTRL), then use the double arrow button to move them both to the right pane. With automatic placement, you can assemble more than one instance of a component at the same time.

Assembling Components using *Replace*

Another useful assembly utility is the *Replace* function. This works particularly well with components defined by family tables. All that is required to replace one component with another is that the assembly references be the same for both. This is guaranteed with a family table and especially when a component interface is defined.

Select one of the bracket bolts on the vertical tubes. In the RMB menu, select (in the **Edit Actions** group) *Replace*. In the **Replace** dialog window, the **Family Table** option is the default. Pick the *Open File* button to the right of the collector under **Select New Component** and choose one of the other members of the family table. Since these have the same component interface, they can be freely substituted in an assembly. Choose *OK* and the new component replaces the old one.

Once all the bolts have been added, the assembly is now completed. Resume everything and make sure everything is visible. In the **Operations** group and **View(Visibility)** groups select:

>   *Resume > Resume All*
>   *Unhide All*

Save the model! There are a few odds and ends missing, such as washers and nuts to put on the bottom of the bolts and some additional components on the axles. You might like to try to add these some time. It would also be convenient either to group all the bolts together, or even better, to put them all on a specially defined layer (use a rule to select entities for the layer).

Performing an Interference Check

Let's see if we have any gross geometric problems with the design of the cart. We want to find out if any of the parts are overlapping or interfering. We shouldn't be too concerned about the bolts, so it will speed things up considerably if we suppress all the bolts out of the assembly. If we had all the bolts on a single layer, we could select them for suppression that way. Instead, open up the model tree and select the bolts at the end of the model tree. Then in the mini toolbar, select *Suppress*.

To check for interference with the remaining components, in the **Analysis** ribbon select

Global Interference

In the window that opens up, select the *Preview* button to compute the interference. If all goes well, after a minute or two you will be informed that there are no interfering parts. (If you haven't defined the left spring as flexible, do that now and rerun the interference check.)

What happens if the wheel suspension moves to a new position? *Regenerate* the assembly with the ground clearance set to 330. Perform the interference analysis again. This time, you should be informed that there is interference. In the **Global Interference** window, a display box shows the pairs of parts involved. Highlighting one of these lines will cause the interfering parts, the spring and the wheel mount plate, to highlight in the graphics window. Zoom in on the interfering area. The interference volume is shown in red. If you expand the Global Interference window to the right, the actual volume of interference is also indicated. There are a number of ways that you could eliminate this interference, and you can do that on your own. What happens if you set a ground clearance of 350? Why? (Scroll back through the message window to find out!) Set the suspension height back to a ground clearance of 250.

Assembly Drawings

The essential commands for constructing drawings were discussed in the first Creo tutorial and in the previous lesson. You will find that, for assemblies, multi-model and multi-sheet drawings are critical to help you organize the task of presenting and detailing the design. As for creating the assembly, it would be a good idea to spend some time planning the layout of all the part drawings first, before committing to any particular scheme. We are not going to do any of that here, since we have already covered many of the basic drawing functions. Also, the cart model could easily involve 20 to 30 drawings for a complete documentation package. We will, however, have a look at a couple of drawing functions that are very commonly used with assemblies.

Before we proceed, **Resume** all the parts in the assembly (**Resume > Resume All**).

One important item that has not been discussed is the creation of a bill of materials (BOM) for the assembly. When you are in assembly mode, you can easily create a BOM by selecting the command in the **Investigate** group:

> **Bill of Materials > Top Level | OK**

The browser window opens showing a listing of all parts in the assembly, broken down by subassemblies and a total part count at the end of the listing. Notice (see the message area) that the BOM file is automatically placed in your working directory with the file name *cart_total.bom*.

What we will do here is to create a BOM on a drawing of the assembly, and attach balloons to some of the components in one subassembly. This is essential design documentation.

Creating a Drawing BOM

Close the browser and model tree and start up a new drawing called *[cart]*. Deselect the option "Use default template," then **OK**. Check the option "Empty with format" on the next window and select the format *tut_format* developed in the last lesson. Complete the prompts as required to specify any parameters in the drawing (like *drawn_by*). Add a general view of the cart assembly using the RMB menu command:

> **General View > No Combined State | OK**

and place the view slightly to the left of center on the drawing sheet. Set the orientation of the view as desired. If you happen to have a favorite view saved in the assembly (like a right isometric), you can select this directly. Use the **Edit > Value** command to change the sheet scale to **0.075**. Using **Parameters** in the **Tools** ribbon, set the **Look In** option to **Assembly** and pick on the cart model. Create or set the value of the description parameter to something like **[Cart Project]**. Close the parameters window. The sheet should look like Figure 47.

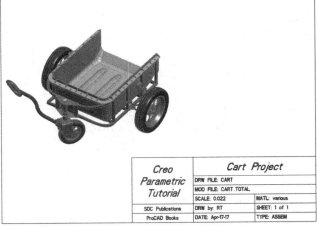

Figure 47 Drawing format and general view

Creating the BOM Table

Our BOM on the drawing will be created using a repeat region defined in a table (see Figure 50 for the final result). Go to the **Table** ribbon; select **Insert Table** in the RMB pop-up in the graphics window. Then pick the second button (*Leftward, Descending*), and set **4** columns, **2** rows, row height **0.25**, column width **0.625**, then *OK*. Locate the table using *Absolute Coordinates* (X = 10.625, Y = 8.125). When the table appears, adjust the width of columns 2 and 3 to **1.5**.

Now create the column headings. Starting in the farthest left column (call it column 1), double click in the cells in the top row and enter the following text:

column 1	**INDEX**
column 2	**Component Name**
column 3	**Component Type**
column 4	**QTY**

Drag a rectangular selection box around the table and in the RMB pop-up, select *Text Style*. Change the Character **Height** to **0.125** (real inches on the sheet), and the **Horizontal** and **Vertical** alignments to *Center* and *Middle*, respectively. Check the changes out with *Apply*, and then select *OK* when you are satisfied. The table should look like Figure 48.

INDEX	Component Name	Component Type	QTY

Figure 48 BOM column headings in table

Creating the Repeat Region

In the **Table** ribbon **Data** group select *Repeat Region*. Then

Add > Simple

and pick on the first and fourth cells in the second row, then select *Done*. Now we add some text into the cells.

Select each cell in the second row in turn. For each, in the RMB pop-up select *Report Parameter* and enter the following:

INDEX	Component Name	Component Type	QTY
rpt.index	asm.mbr.name	asm.mbr.type	rpt.qty

Figure 49 Repeat region parameters

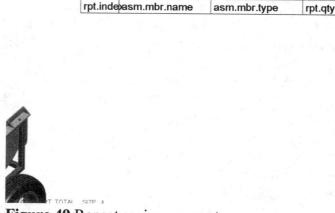

column 1	**rpt.index**
column 2	**asm.mbr.name**
column 3	**asm.mbr.type**
column 4	**rpt.qty**

These text strings will appear in the table, overlapping the cells a bit. Don't worry about that now. Adjust the text style as we did for the column headers, except that set **Horizontal(Left)**. See Figure 49.

Creating the BOM

We now have everything defined. In the Data group, select *Update Tables* (or use the RMB pop-up). The table will expand down considerably off the sheet. The table contains a large number of duplicated entities (especially the bolts). We'll combine these by selecting *Repeat Region* (in the **Data** group), then select

Attributes > [pick on the region] > *No Duplicates | .. | Done/Return*

The table now collapses and will appear as shown in Figure 50. This is the BOM for the top level assembly. It identifies which components are parts and which are subassemblies, and how many components are involved. In the **TBL REGIONS** menu, select *Done*.

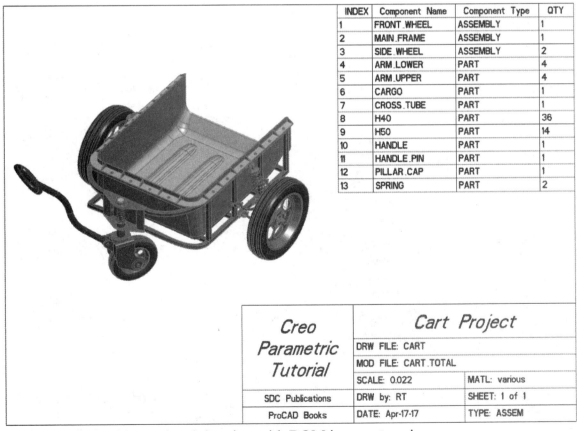

INDEX	Component Name	Component Type	QTY
1	FRONT WHEEL	ASSEMBLY	1
2	MAIN FRAME	ASSEMBLY	1
3	SIDE WHEEL	ASSEMBLY	2
4	ARM LOWER	PART	4
5	ARM UPPER	PART	4
6	CARGO	PART	1
7	CROSS TUBE	PART	1
8	H40	PART	36
9	H50	PART	14
10	HANDLE	PART	1
11	HANDLE PIN	PART	1
12	PILLAR CAP	PART	1
13	SPRING	PART	2

Creo Parametric Tutorial	Cart Project	
	DRW FILE: CART	
	MOD FILE: CART TOTAL	
	SCALE: 0.022	MATL: various
SDC Publications	DRW by: RT	SHEET: 1 of 1
ProCAD Books	DATE: Apr-17-17	TYPE: ASSEM

Figure 50 Assembly top level drawing with BOM in repeat region

Adding Balloons

Add another drawing model, for example the side wheel subassembly. Set this as the active model and then add another drawing sheet. If you recall, doing things in this order means that the drawing title block will automatically refer to the active model. Methods to do this were discussed in the previous lesson. On the second sheet, create an exploded view with an appropriate scale. This is most easily done if the explode distances are defined in assembly mode and a named view created there, as suggested earlier. Go to the **Table** tab in the ribbon and create another BOM for the subassembly using the same procedure and repeat region parameters as on sheet 1, and remove duplicates. The QTY may not appear until you have removed duplicates. Use *Format > Text Style* to center the text in the cells.

Finally, in the **Balloons** group select (and watch the message line)

Create Balloons - By View

Pick on the repeat region, then pick on the exploded view. Highlight each balloon to drag it to a better position. You can use other RMB menu commands to access other options for positioning and formatting of balloons. The final drawing should look something like Figure 51. Save the drawing.

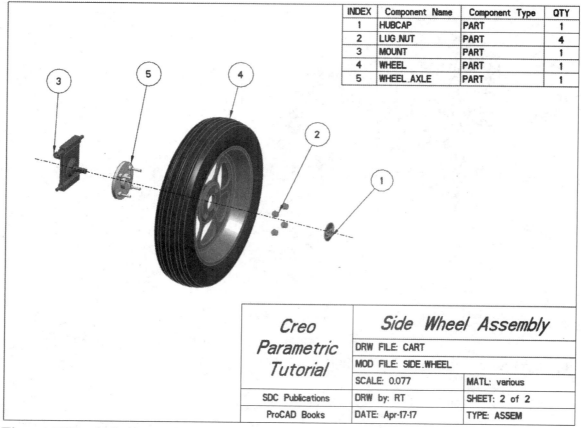

Figure 51 Exploded assembly view with BOM and balloons

There is considerably more you can do with the BOM. You can, for example, attach a cost to each component and sum the cost for all entries in the table. You can add **Bulk Items** such as paint, solder, glue, and so on. These are actually created as components in the top-level assembly. The repeat regions can also be nested so that the entire BOM contains components sorted by subassembly.

Conclusion

Well, this is the end of this tutorial. There are some exercises at the end of this lesson to give you some ideas for extending the cart project. Hopefully, the tutorial has shown you a number of new tools and techniques to allow you to take on more challenging and complex modeling projects, and to do them more efficiently. As always, there is lots more to learn about how the basic Creo modeling tools can be applied. Only with practice and experience will you become very adept at this task. You are certain to come across unique and tricky modeling tasks, and the more acquainted you are with the tools, the better you will able to succeed. You should be developing a feel for how these modeling tools can be employed to produce clean, flexible, and easy to use models, and how advanced modeling differs from simple geometry creation.

Questions for Review

1. What must you consider when planning to create a component pattern?
2. Where can you find the merge command?
3. What is the order of component selection when using *Merge*?
4. What are Boolean operations?
5. What happens to the assembly file used to create a mirrored part?
6. How do you select the constraints to be used to repeat the placement of a component?
7. What restriction do you have to keep in mind when using *Reference Pattern* to place new components?
8. What was the purpose of the assembly datum curve in the side frame subassemblies?
9. What is a component interface and what can it do? How do you set up a component interface?
10. How do you define a flexible component?
11. What happens to the file stored on disk when a component has a flexible dimension?
12. Can a component be declared flexible when you are still in PART mode?
13. What are the shortcut procedures available to allow fast/efficient assembly of multiple instances of the same component? Which of these involve planning ahead?
14. What happens if you edit a dimension of a family table instance in the assembly if (a) the dimension is driven by the family table? and (b) the dimension is not driven by the family table?
15. How do you determine if there are interfering parts in the assembly?
16. When you are in assembly mode, how do you create a bill of materials?
17. How do you get rid of repeated entries in a BOM on a drawing?
18. What happens if a defined component interface does not yield a fully constrained component placement in the assembly?
19. Can a subassembly be made flexible?
20. If a flexible component occurs (with different dimensions) at several places in an assembly, how many entries appear in the BOM?
21. Does a flexible dimension have to refer to solid geometry? Could it be used to locate a datum?
22. Suppose you want an assembly BOM to appear automatically on a drawing. Do you add the table and repeat regions to the sheet format or a drawing template?

Some Final Project Exercises

1. Modify the cargo bin so that it will dump. Perhaps you could add a spring assisted lift for this.
2. Set up the model so that the location of the side wheels (front to back) is adjustable.
3. Determine the total weight of the cart and find its center of gravity - if it's behind the wheels we're in big trouble.
4. Design a waterproof canopy for the cart.
5. Design an easily removed tailgate.
6. Add parameters and relations to automatically change the lengths of the side arms on the suspension.
7. Add parameters and relations to change the size of the cargo bin and main frame.